Key Thinkers for the Information Society

Does the contemporary 'information society', ushered in by the widespread use of new information and communication technologies like the Internet, require new ways of thinking to understand it? Should we forget all that has been written previously about the interaction of society and technology? *Key Thinkers for the Information Society* introduces eight writers from the past, each of whom has something important to tell us about the 'new age'.

The writers selected offer valuable insights into the information society, but their work is often neglected by contemporary commentators. Individual chapters introduce the work of **Walter Benjamin**, **Murray Edelman**, **Jacques Ellul**, **Harold Innis**, **Lewis Mumford**, **Karl Polanyi**, **Elmer Eric Schattschneider** and **Raymond Williams**. Each chapter provides key biographical information, places the writer in the context of his discipline, and demonstrates why his analysis is useful for understanding the role that ICTs play today.

Key Thinkers for the Information Society demonstrates that our 'brave new world' has significant continuities with the past and offers fresh perspectives on contemporary debates.

Christopher May is Senior Lecturer in International Political Economy, University of the West of England.

Key Thinkers for the Information Society

Edited by Christopher May

in association with
British International Studies Association,
International Communications Working Group

Routledge
Taylor & Francis Group

LONDON AND NEW YORK

First published 2003
by Routledge
11 New Fetter Lane, London EC4P 4EE

Simultaneously published in the USA and Canada
by Routledge
29 West 35th Street, New York, NY 10001

Routledge is an imprint of the Taylor & Francis Group

© 2003 Christopher May, selection and editorial matter;
individual chapters, the contributors

Typeset in Times by
Keystroke, Jacaranda Lodge, Wolverhampton

Printed and bound in Great Britain by
TJ International Ltd, Padstow, Cornwall

British Library Cataloguing in Publication Data
A catalogue record for this book is available
from the British Library

Library of Congress Cataloging in Publication Data
Key thinkers for the information society/edited by Christopher May.
        p. cm.
    Includes bibliographical references and index.
    ISBN 0–415–29672–2 (hb)—ISBN 0–415–29673–0 (pb)
    1. Information society. 2. Information technology.
    3. Communication—Technological innovations.
    I. May, Christopher, 1960–

    HM851 .K49 2002
    303.48′33—dc21   2002027524

ISBN 0–415–29672–2 (Hbk)
     0–415–29673–0 (Pbk)

Contents

Figures and tables

Figures

Tables

Notes on contributors

Robin Brown is the Head of the Institute of Communications Studies, University of Leeds, and a founding member of the British International Studies Association, International Communications working group.

Andrew Chadwick is a lecturer in Politics at Royal Holloway, University of London.

Edward Comor is Assistant Professor in International Communication at American University's School of International Service in Washington, DC and was the founding Chair of the International Communications Section of the International Studies Association.

Marianne Franklin teaches International Political Economy and International Communications at the University of Amsterdam and is a founding member of the British International Studies Association, International Communications working group.

Des Freedman is a lecturer in Communications and Cultural Studies at Goldsmith's College, University of London.

Karim H. Karim is Assistant Professor at Carelton University's School of Journalism and Communication in Ottawa.

Christopher May is Senior Lecturer in International Political Economy at the University of the West of England (Bristol), and a founding member of the British International Studies Association, International Communications working group.

Kenneth S. Rogerson is a lecturer at the DeWitt Wallace Centre for Communications and Journalism, Duke University (Durham, NC), and a founding member of the International Communications Section of the International Studies Association.

Acknowledgements

This volume is a collection of articles that were originally published as part of an ongoing series on analysts whose work is relevant to the study of the information society but who were seldom cited in such discussions. The series is edited by Christopher May on behalf of the British International Studies Association, International Communications working group (information on the group can be obtained from Jayne Rodgers: jayne@ics-server.novell.leeds.ac.uk). The editor welcomes further contributions to the series from scholars outside the working group, and proposals should be forwarded to him either by email (christopher.may@ uwe.ac.uk) or by post to:

Christopher May
Faculty of Humanities, Languages and Social Sciences
University of the West of England
Coldharbour Lane
Bristol BS16 1QY, UK

The editor and the working group thank the editors of *Information, Communication and Society*, and the publishers, Taylor & Francis for permission to reprint the following:

The chapter on Walter Benjamin was specially prepared for this volume. A substantially different article will appear in *Information, Communication and Society* in the future.

The chapter on Murray Edelman originally appeared as:
Andrew Chadwick (2001) 'The electronic face of government in the Internet age: borrowing from Murray Edelman', *Information, Communication and Society* 4(3): 435–457.

The chapter on Jacques Ellul originally appeared as:
Karim H. Karim (2001) 'Cyber-Utopia and the myth of Paradise: using Jacques Ellul's work on propaganda to analyse information society rhetoric', *Information, Communication and Society* 4(1): 113–134.

The chapter on Harold Innis originally appeared as:
Edward Comor (2001) 'Harold Innis and the "Bias of Communication"', *Information, Communication and Society* 4(2): 274–294.

The chapter on Lewis Mumford originally appeared as:
Christopher May (2000) 'The information society as mega-machine: the continuing relevance of Lewis Mumford', *Information, Communication and Society* 3(2): 241–265.

The chapter on Karl Polanyi originally appeared as:
Ken Rogerson (2002) 'Addressing the negative consequences of the information age: lessons from Karl Polanyi and the industrial revolution', *Information, Communication and Society* 6(1): (in press).

The chapter on Eric Elmer Schattschneider originally appeared as:
Robin Brown (2002) 'The contagiousness of conflict: E.E. Schattschneider as a theorist of the information society', *Information, Communication and Society* 5(2): 258–275.

The chapter on Raymond Williams originally appeared as:
Des Freedman (2002) 'A "technological idiot"? Raymond Williams and communications technology', *Information, Communication and Society* 5(3): (in press).

Chapter 1

Editor's introduction

Christopher May

Many people, in policy-making circles, in the press, in universities, claim we have entered a new age, governed by a 'new paradigm' where society and its economic relations are no longer primarily organised on the basis of material goods. These writers and commentators often suggest that the accelerated flows of information and the increased utilisation of knowledge have fundamentally transformed society. They point to the decline of manufacturing compared to the prospering information-rich service sector as exemplary of such a transformation. This is sometimes referred to as the arrival of a (global) information society, or discussed as a 'weightless world' or, following Manuel Castells, the arrival of a new network society. We are told, in the catch-all shorthand of the new millennium, that 'the Internet' is changing everything: new information and communication technologies (ICTs) have produced a revolution, a remaking of the world. Furthermore, all that we previously knew is useless for thinking about this new world; *the new age requires new thinking*. Old ideas must be dispensed with, we must welcome the new information society into our lives and embrace this new age. We must look to the future and forget the past.

Certainly, ICTs and their networking capacity (most readily manifest on the Internet) are having considerable social, political and economic effects. But, these changes are frequently overdrawn, leaving continuities with the past wilfully obscured at the same time that the new age is presented as unprecedented (May 2002). In direct contrast to the hyperbole which surrounds these claims of revolution, this book starts from the position that there is an urgent need to reintegrate the information society and its technological apparatus into an understanding of the continuing history of technology and society. The contributors to this volume share the view that there are sufficient analytical tools to hand, without the continual invention of new paradigms, to understand the current stage of technological advance. Crucially, we do not suppose that contemporary developments are completely novel. There *have* been changes in the forms of social relations and the technological practices which help shape our lives, but these are not so profound as to render obsolete previous approaches to understanding the role of information, technology and information technologies in society.

Our premise is that there is no need to discard all previous analysis and insight, no need to dispense with our previous 'ideological baggage'. The claim that we *should* is, of course, ideological in itself. It represents a dismissal of well-developed arguments regarding the contested and contingent character (and practices) of modern capitalist society, while also presenting a specific set of contemporary social relations as natural and outside history. As Carolyn Marvin remarked in an earlier period of revolutionary fervour: 'Information age rhetoric is the start-fresh propaganda of our age' (Marvin 1987: 61). Much contemporary information society commentary is also technological determinist (naturalising the thrust of technology). It disavows our role in applying the technologies that are available to us, and denies the social shaping of particular technological advances and innovations. However, new technologies do not represent neutral mechanisms, nor novel developments lacking histories, but rather are a technological manifestation of the historically specific social system in which they have emerged.

Nevertheless, many of those who proclaim that the development and deployment of new ICTs have prompted a 'new age' start from the premise that technological changes bring in their wake major shifts in the societies which use these transformed technologies. Indeed, to argue that ICTs are producing a new age *requires* this assumption, because otherwise it would make little sense to focus so keenly on ICTs. Technological determinists think (to varying degrees) that technological advances happen automatically; they see a logic to technological developments which is beyond our influence. Technological advances sweep all before them, leaving us struggling to accommodate or understand their implications.

As Langdon Winner noted in the late 1970s, this perception of autonomous technological development has frequently become the fear of 'technology out-of-control'. The identification of a technological imperative is the combination of the recognition of actual processes of change in technologies and the suggestion that humankind's common disposition is to react and accommodate change, not to try to reverse or redirect it. Linked together 'the process and the disposition create what can be called technological dynamism, a forceful movement in history which continues largely without human guidance' (Winner 1978: 105). By linking a recognition of changes in technology with an assumption of human acquiescence in the face of technologies' logic, the role of technological change in history is magnified into determinism. Taking changes in technology as the most important single factor in explaining any particular shift in society, technological determinists deny (or ignore) the role of social and political choice, thereby obscuring the social embeddedness of technology. And for the contributors to this book, most importantly such determinism denies the continuing relevance of our historical understanding and analysis of society, as well as the alternatives such views represent.

The notion of an 'information age' underlies most writing on the transformation wrought by ICTs. As used by the historian Eric Hobsbawm for instance, previous ages (the ages of revolution, capital and empire) gave way to the 'age of extremes' in the twentieth century (Hobsbawm 1994). As Jennifer Slack noted:

> An age connotes an all pervasive logic, a logic that requires that everything be explained in its own terms . . . The information age thus hails all subjects as trapped in its logic.
>
> (Slack 1984: 253)

And therefore, the use of the term 'age' tells us something about what is intended by its proponents. There has been no shortage of assertions of a new age ushered in by the widespread use of powerful new ICTs, an age which has seen profound changes in the logic of social existence. Writers from Alvin Toffler in the 1970s and 1980s, to more recent commentators like Manuel Castells or Don Tapscott, have all claimed that this new age has already arrived (May 2002: 3–14). The hope that we might have some (political and/or social) control over the direction of the information revolution, while sometimes evident in earlier commentary, has since faded. Daniel Bell, Theodore Roszak and others writing in the 1960s and 1970s thought that new technologies might prompt considerable social engagement, and would be shaped to maximise their social and welfare potential. More recently the information society has become a wave which we can surf but cannot change or modify. This shift from engagement to passive accommodation has been accomplished by presenting these developments as epochal rather than merely taking place *within* contemporary society.

One of the most influential treatments of the epochal story was outlined by Alvin Toffler, who identified the 'information age' as the third wave (Toffler 1980). Indeed, the 'third wave' continues to find its way into much that is written about the information society, and seems to have attained the position of 'received truth', even if its author is often unacknowledged. Essentially Toffler's argument is that there have been two previous (universal) technological revolutions, or waves, in the way humans have organised their socio-economic affairs (the agricultural and industrial), and now the information revolution is a third (May 2002: 21–23). The deployment and use of increasingly powerful ICTs is the key driver of this third wave in the same way that emergent farming and industrial technologies spurred the previous two waves or revolutions. As we emerge from the transitional period between the second and third waves, society will have been remade. This remaking is reflected in the way states interact (Toffler and Toffler 1993), in the way society is organised and the sorts of economic activities which will be valued and provide employment (Toffler 1980). Whether explicitly drawn from Toffler, or not, this analysis implicitly combines the notion of 'creative destruction' originally developed by Joseph Schumpeter to describe the way capitalism moved forward by rendering redundant previous technologies, and Nikolai Kondratieff's argument that the global economy moved through 'long waves' of economic development. Joining these ideas together produces a depiction of periods of upheaval (related to technological change) between long waves.

This leads writers like Nicholas Negroponte to argue that the information revolution is a natural force which brokers no resistance; there is no alternative to *being digital* (Negroponte 1995). The character of the world is being transformed, not by social or political negotiation but by technology. The information age becomes the explicit expression of inevitability, of progress, of the technological future made present. But even if we accept that new ICTs are having *some* effects on society, there is no need to also assume that these developments are recent in origin; no need to assume these changes reflect a new historical epoch.

Some writers have started to trace the origins of the information age back to at least the nineteenth century if not before (Chandler and Cortada 2000; Levinson 1997; Winston 1998). This might lead us to agree with William Wolman and Anna Colamosca that 'by endowing libraries across the country, Andrew Carnegie created an earlier knowledge revolution in the United States whose scope at least matches that of the information revolution created by Bill Gates and his competitors' (Wolman and Colamosca 1997: 75). Or looking further back, we can recognise that in many ways the informational chaos of the Internet is strikingly similar to the problems encountered by readers in the sixteenth and seventeenth centuries who were frequently unable to establish the veracity of printed texts, because of problems of piracy and unauthorised (amended and edited) editions (Johns 1998: 171). By widening our perception of what information and communication technologies are, which is to say avoiding the contemporary fixation with computers and the Internet, we can easily recognise a much longer process at work.

By taking the longer view of the development of the information society therefore, we can reject the claims for contemporary transformation, without suggesting that nothing has changed. In the early 1980s Wilson Dizard noted, that the

> so-called communications revolution is, in reality, a succession of three overlapping technological stages that have taken place during the past one hundred and fifty years. The first of these was the Wire Age (1844–1900), the second was the Wireless Age (1900–1970), and the third is the one we are now entering – the Integrated Grid Age, in which wire and wireless technology are brought together in powerful combinations which will form the structure of the future global information utility.
>
> (Dizard 1982: 47)

This longer time line is also supported by James Beniger, who suggested that the emergence of an information society was predicated on the 'control revolution', itself a response to the forces unleashed by the industrial revolution (Beniger 1986). At the centre of the 'control revolution' was the ability of instantaneous feedback to allow continuous adjustment and the enhancement of material usage. Beniger makes explicit the links between mechanical control systems and the rise of the information society while stressing the roots of such developments in the nineteenth century and earlier. Recognising the importance of the manipulation of information for the management of industrial organisation and production, and focusing on the increasing ability of technology to allow the control of processes, does not involve a claim that this is only a contemporary phenomenon.

Furthermore, the celebration of the possibilities of the Internet as part of the new information age is hardly original. In a study which makes explicit the parallel developments of telegraphy and the Internet, Tom Standage identifies some striking similarities in public responses to both technologies. In both cases public reaction was 'a confused mixture of hype and scepticism' (Standage 1998: 194). Like many recent writers, Victorians celebrated the telegraph as a new mode of communication that would further democracy and social communication for the good of all (to bring about peace among nations) at the same time as they saw it prompting new methods of control. Businesses could be centralised (control could be direct with less delegation to outlying plants), and governments could much more easily direct their armies (and their societies) through swift and authoritative communication. As Standage concludes:

> Today, we are repeatedly told that we are in the midst of a communications revolution. But the electric telegraph was, in many ways, far more disconcerting for the inhabitants of the time than today's advances are for us. If any generation has the right to claim that it bore the full bewildering, world shrinking brunt of such a revolution, it is not us – it is our nineteenth-century forebears.
>
> (Standage 1998: 199–200)

The celebration of the information society, the celebration of the new age, is predicated on the novelty of today. However, this 'new age' is neither unprecedented, nor necessarily as novel as often presumed.

This leads to the conclusion that the argument for a new age, to which 'there is no alternative', is a profoundly *political* argument. As Ziauddin Sardar notes:

> in a very subtle way, predictions and forecasts silence debate and discussion . . . We are locked in a linear, one-dimensional trajectory that has actually foreclosed the future . . . There is no such thing as *the* future; there are many, many futures. And our concern should be with what the future ought to be, what we want it to be.
>
> (Sardar 1999: 27)

We need to question the assertions that are frequently made about the use, value and deployment of ICTs. We need to be clear why certain developments are presented as inevitable and why others are regarded as non-sensical. We need to carefully examine these assumptions and challenge them if needs be. And to do this we need the intellectual armoury that has been developed over many years to question and investigate modern society, not dispense with it on the grounds it has become suddenly obsolete.

As before (information) society will be what we make of it, and therefore the battle to conceive of the future is important and should not be left to the technicists, or the policy makers and especially not to the information society celebrants. There is nothing natural, nothing inevitable about the information society: while we can only make our own history in the circumstances in which we find ourselves, we should recognise that these circumstances are not as fixed nor as narrow as many commentators on the information society tell us. That said, although it obscures much, the notion of the information society is not completely without foundation. There *have* been changes, but they are not as great in magnitude as supposed, nor necessarily in the directions forecast. There is a disjuncture between the claims for the information society and its readily identifiable manifestation. To understand our contemporary predicament we need to go back to those who have thought about society prior to the onset of the so-called 'information age', to recover insights from their work that remain relevant to today.

Retrieving insights from the past

There have been three waves of comment on the information society: an American phase (1960s to 1970s), a modernisation phase (late 1970s to early 1980s) and a global (or Internet) phase (from the mid-1990s to now) (May 2002: 3–12). However, what is particularly notable about this most recent wave of comment is its disconnection from the critiques which were advanced in the previous two waves; critiques which were related, as is this book, to social analysis from before the 'revolution' (see for example Heyer 1988; Kuhns 1971). The following chapters reconnect the information society with its (and society's) past. They establish the importance of revisiting past criticisms of the role of infor-mation and technology in society, when thinking about information society today.

The choice of the writers featured in this volume reflects work being carried out in the International Communications working group of the British International Studies Association, and the International Communications Section of the International Studies Association in the United States. Members of both groups have nominated writers who they believe not only offer analytical insights regarding the information society, but also have too often been obscured by the contemporary clarion call for novel analysis. While the following chapters represent the choices of the seven contributors (and myself), neither I nor they are suggesting that these are the only authors one might (re)investigate. The continuing series of articles in *Information, Communication and Society* will explore further 'key thinkers' in the future. However, for each contributor to this volume, their chosen writer has something important to offer which is underplayed or absent from much current analysis of the information society. Each chapter makes the case for the re-examination of a particular view of themes that are important for understanding the role of ICTs in today's society. The chosen analysts come from diverse perspectives reflecting the wide multidisciplinarity of information society studies, and the range of accounts of the information society which can be developed.

Each of the chapters stands alone, but they also represent a collective critique of the assertion that the global information society characterises the new millennium. Drawing together the insights and 'rescued' analyses that the contributors have presented in their chapters allows us to (re)establish a more nuanced, less glib and less passive analysis of the 'information age'. As I have already suggested, one of the key problems that much commentary celebrating the arrival of the new age succumbs to is the casual acceptance of an implicit (or sometimes quite explicit) technological determinism. As Des Freedman discusses in his chapter on Raymond Williams, this is hardly the first time that the celebration of a new technology has suffered from this shortcoming. As Freedman makes clear, Raymond Williams developed an important critique of determinism as related to communications (and other) technologies before the arrival of the latest technological revolution.

In arguments that will be broadly familiar from the first section of this introduction, Williams questioned the central assumptions of technological determinism. Contrary to the manner in which technological change is presented as an independent force (or 'independent variable'), Williams stressed that technologies are embedded in the practices and social relations of the societies in which they appear. Although developed in a different manner, there are clear parallels here with Lewis Mumford's use of the notion of *technics* rather than technology to discuss the historical development of human technological prowess. This recognition of embeddedness led both Williams and Mumford to emphasise the human choices that produce the dynamics and direction of technological development. While Mumford took a long-term overview of this issue (emphasising the interplay between authoritarian and democratic technics across human history), Williams focused on modern communications technologies to make the same point. We are not slaves to new technology, rather we have (or at least some of us) have chosen the technological solutions that have been developed, and these solutions represent particular interests. Thus, as Freedman's reading of Williams emphasises, the notion of the inevitability of the direction of technological change is not the result of the logic of any particular technology, but rather is the reflection of the dominant groups' needs in society. These powerful groups (often successfully) attempt to reify or naturalise their own interests.

In Marianne Franklin's close reading of Walter Benjamin's famous essay 'The work of art in the age of mechanical reproduction', this issue of the presentation of inevitability is again criticised, but from a different direction. For Benjamin, as Franklin makes clear, the notion that technological changes produce specific shifts in artistic representations was too simplistic and acted to limit the possibilities of human expression. Rather than the pessimistic critical theory of the Frankfurt School, Benjamin suggested that it is possible (although not inevitable) that the possibilities and potentialities of new technology can be wrested from the control of the powerful interests who seek to shape them to benefit themselves. Here again, the central issue of indeterminacy is foregrounded. Indeed, as both Andrew Chadwick in his retrieval of Murray Edelman's approach to symbolic politics, and Karim H. Karim in his exploration of Jacques Ellul's writings on myth and propaganda, point out, it is this very indeterminacy that powerful interests try to deny and foreclose. By developing a particular story of the information age, other narratives, other possibilities are hidden or rendered non-sensical.

Chadwick notes that utilising Edelman's perspective on the use of symbolism in politics helps reveal the way that the Internet and other ICTs are being deployed to present a very specific view of the information society and its possibilities. This closing down of the potentialities of new ICTs reflects the struggle over their social use and suggests that technologies have no natural logic which indicates how they will be used in the contemporary structures of society. The myths of technological advance (linked, as they are, to ideas of utopia and paradise) have been central to the political discourse of technology for centuries. For Karim therefore, Ellul's discussion of propaganda is a useful tool for opening up this discourse to reveal its partiality and the mobilisation of political interest it represents. The propaganda which surrounds the notion of the information society or the new 'information age' is used to reinforce the myth of inevitability. The symbols of information society generate a mythical future which is then turned into the inevitable endpoint of technological advance. The information society becomes a self-fulfilling prophecy, demanding actions to deal with an impending social change, actions themselves which *produce* this social change.

However, as I argue in my chapter, Lewis Mumford suggests that this heated discussion over the direction of technological change is hardly novel. Rather the yoking of technology to specific social and political ends has been the way technology itself has advanced; innovations and new uses have been a direct response to the previous technological deployments and their contested effects. As the possibilities of new technologies have been developed, so their use and deployment have been challenged and subverted. In a very real sense, there can be the 'double movement' which Kenneth S. Rogerson identifies in his chapter on Karl Polanyi. Nevertheless, Robin Brown's discussion of Elmer Eric Schattschneider reminds us that politics is still about power, and while the expanded flows of information may have an impressive impact on the practices of politics, this does not invalidate nor make obsolete the previous patterns of political interest.

Widening flows of information may well enhance the visibility and profile of particular pressure groups and campaigns, but this is not enough to transform politics. The political potential of such shifts in the political milieu can be carried forward only through political action; information is not sufficient for political change. Information, again as Polanyi would testify, is embedded in social relations; information may well be a 'fictitious commodity' not necessarily subject to the thrust of capitalist socio-economic relations, but there are powerful forces arrayed to (re)produce such an arrangement. 'Indeed, the tendency to bias which Harold Innis made central to his treatment of communication, while clearly being disturbed by new technologies, is relatively intransigent. As Edward Comor stresses, whatever the claims regarding the new information politics, ICTs have been developed and deployed within a capitalist system, that (re)produces a particular bias regarding their possibilities and potentialities. Thus too often the technological issues and problems to which politics attends are are not those which some of us might regard as crucial for a better future.' Thus, despite the claims for a new information politics much remains remarkably familiar, a point underlined by Chadwick's deployment of Edelman's work in his discussion of the emergence of 'e-government'.

All the analyses in their own way mirror Williams' explicit critique of technological determinism. Each writer comes at the issue from a different direction, emphasising different effects of the myth of inevitable tech-nological development. The critical voices discussed in the following chapters remind us that technology is indeterminate, it has no fixed logic, no pre-inscribed trajectory. The information age, like previous techno-logical 'ages', will not be unitary or uncontested in character; society has a profound impact on ICTs every bit as much as they have an effect on society. This is not to suggest a voluntaristic notion of technological advance. Rather, the use and deployment of (information) technology is deeply embedded in the existing social, political and economic practices and relations of society. As all the contributors discuss in various ways, the attempt to present the information society as a teleology is itself a myth, an ideology, deployed to shape the future and deny other alter-natives. This, I would emphasise, is not to suggest that nothing is changing, it is not to claim that ICTs are having no effect on our lives or society. However, it is to argue that we need to integrate our understanding of these effects with social analyses that have been developed in the past. All of the contributors hope that you will find something of use and interest in their presentation of these writers' work, something that will enable you to develop your own ongoing engagement with the portrayal of the information society. None of these chapters is a substitute for going to the original works, and all of us hope that once you have discovered the analyst(s) that reflect your own interests best, you will delve into their works for further insights and analytical resources.

References

Beniger, J.R. (1986) *The Control Revolution: Technological and Economic Origins of the Information Society*, Cambridge, MA: Harvard University Press.

Chandler, A.D. and Cortada, J.W. (eds) (2000) *A Nation Transformed by Information*, New York: Oxford University Press.

Dizard, W.P. (1982) *The Coming Information Age: An Overview of Technology, Economics and Politics*, New York: Longman.

Heyer, P. (1988) *Communications and History: Theories of Media, Knowledge and Civilisation*, Westport, CT: Greenwood Press.

Hobsbawm, E. (1994) *Age of Extremes: The Short Twentieth Century 1914–1991*, London: Michael Joseph.

Johns, A. (1998) *The Nature of the Book: Print and Knowledge in the Making*, Chicago: University of Chicago Press.

Kuhns, W. (1971) *The Post-Industrial Prophets: Interpretations of Technology*, New York: Weybright & Talley.

Levinson, P. (1997) *The Soft Edge: A Natural History and Future of the Information Revolution*, London: Routledge.

Marvin, C. (1987) 'Information and history' in J.D. Slack and F. Fejes (eds) *The Ideology of the Information Age*, Norwood, NJ: Ablex.

May, C. (2002) *The Information Society: A Sceptical View*, Cambridge: Polity Press.

Negroponte, N. (1995) *Being Digital*, London: Coronet/Hodder & Stoughton.

Sardar, Z. (1999) 'The future is ours to change', *New Statesman* 19 March: 25–27.

Slack, J.D. (1984) 'The information revolution as ideology', *Media, Culture and Society* 6: 247–256.

Standage, T. (1998) *The Victorian Internet*, London: Weidenfeld and Nicolson.

Toffler, A. (1980) *The Third Wave*, London: Collins.

Toffler, A. and Toffler, H. (1993) *War and Anti-War: Survival at the Dawn of the 21st Century*, Boston, MA: Little, Brown.

Winner, L. (1978) *Autonomous Technology: Technics-out-of-control as a Theme in Political Thought*, Cambridge, MA: MIT Press.

Winston, B. (1998) *Media Technology and Society – A History: From the Telegraph to the Internet*, London: Routledge.

Wolman, W. and Colamosca, A. (1997) *The Judas Economy: The Triumph of Capital and the Betrayal of Work*, Reading, MA: Addison-Wesley.

Chapter 2

Walter Benjamin

Marianne Franklin

> During long periods of history, the mode of human sense perception changes with humanity's entire mode of existence. The manner in which human sense perception is organised, the medium in which it is accomplished, is determined not only by nature but by historical circumstances as well.
>
> (Benjamin 1973: 216)

> [T]aking responsibility for the social relations of science and technology means refusing an anti-science metaphysics, a demonology of technology, and so means embracing the skilful task of reconstructing the boundaries of daily life, in partial connections with others, in communications with all of our parts.
>
> (Haraway 1990: 223)

All manner of financial transactions, manufacturing processes, and service industries would be hard put to function these days without information and communication technologies (ICTs).[1] But the jury is still out on whether they are a good or bad thing for sociocultural and political economic life. Do ICTs enhance or obstruct political practice in liberal democracies? What exactly is their impact on social and cultural relations? Do they improve or undermine cultural and artistic expression? Do they make people more or less sociable, economic growth or institutions more stable or more vulnerable? Meanwhile, ICTs are becoming increasingly part of the arts; prevalent in experimental and popular film, classical and popular music, computer and video games, the (multi) media and entertainment industries, and so on. And techniques of digitalised (re)production mean, in turn, their wide dissemination through the Internet (email and the World Wide Web). Art galleries post their permanent collections and upcoming exhibitions on the World Wide Web; photos and other sorts of images are sent to family and friends through email or posted on personal and institutional homepages by way of scanners, webcams and digital cameras. These can all be accessed, downloaded and distributed by anyone who has the equipment, access, time and basic know-how.

Their own recent digitalisation notwithstanding, photography and film have become well-established genres in both the 'fine arts' and 'popular culture'.[2] But this has not always been the case. Like television, video, computer games and now the Internet/World Wide Web, their swift popular success along with their incorporation into the communication apparatus of left-wing (Socialist/Communist) and right-wing (Fascist) political movements were not welcomed with open arms by the cultural, intellectual or political establishments of the time. Then, as is the case now, commentators were split along political ideological lines and aesthetic sensibilities. Even more reason, then, to revisit an essay that tackles head-on the way politics, art, and technology interact: 'The work of art in the age of mechanical reproduction' (1935–1939) by Walter Benjamin.[3] This essay, a minor *cause célèbre* in European Marxist/left intellectual circles at the time, has become a cult classic for their successors as well as a permanent fixture on philosophy, cultural studies, and art history literature lists.

An enormous intellectual industry has developed around this 'unclassifiable' writer and literary critic (Arendt 1973: 9), critical thinker (Ferris 1996: 1–3), 'non-conformist' Marxist (Leslie 2000: viii) and 'undeniably . . . difficult figure' (Roberts 1982: 3). This chapter is not an attempt to claim Walter Benjamin as a theorist of the 'Information Age' nor assumes responsibility for providing a 'Rough Guide' to his complex and eclectic thought. More than a few in-depth studies, biographies and edited volumes on Benjamin-related themes already exist in any case.[4] Rather, this chapter reads 'The work of art in the age of mechanical reproduction' with respect to both its own techno-historical timeframe and intellectual backdrop and its resonance with latter-day commentaries on the (lack of) aesthetic value, (negative) socio-cultural and/or political economic ramifications of ICTs in general, and the popularity of the Internet/World Wide Web in particular.

The chapter is divided into two main sections. The first two subsections put Benjamin's life and work into context by drawing out some important intellectual, historical and philosophical issues surrounding the writing and publication of this essay. The third subsection sets out some parameters for reading Benjamin in these 'digital days' that follow from these issues; issues that still exert a powerful influence on successive, mainly negative, critiques of techno-cultural change from within Marxian scholarship. The final subsection looks more closely at 'The work of art in the age of mechanical reproduction' vis-à-vis digital reproduction. The second main section extends the field of vision by juxtaposing Benjamin's essay with an equally controversial one written by Donna Haraway. Her 'Manifesto for cyborgs: science, technology, and socialist feminism in the 1980s' (1990[1985]) has become a cult classic for the contemporary 'Age of Digital Production'.[5] These two works are placed in a conversation in order to highlight the way both are interested in the emancipatory potential of ICTs while recognising the flipside: how mechanical (or digital) reproduction becomes enmeshed in (exploitative) social relations of capitalist production. In that respect, Benjamin and Haraway share a Marxian approach that incorporates 'both the factuality of the objective world and its contents and the actuality of subjective human interaction with that objective world' (Leslie 2000: xii). The chapter concludes with a brief discussion on the implications of these readings for (re-)thinking the information age; however defined. Following Benjamin's and Haraway's cues, I argue for critical analyses of technology, the arts, and politics – and the relationship between them – that do not rely on preordained hierarchies aesthetic, political, or social value (Benjamin 1973: 217–218; Haraway 1990: 223).

Walter Benjamin's life in historical context

> The desire to avoid unseemly controversy or over-hasty interpretation has not had very desirable consequences for the understanding of Benjamin, who is probably now even more obscure and misunderstood a figure than he ever was . . . [But] contrary to what many commentators seem to think, the essence of Benjamin's work is that it is not sceptical or pessimistic. It is addressed to the possibilities of immediate implementation, and the realities of practical politics.
>
> (Roberts 1982: 3, 6–7)

Walter Benjamin (1892–1940) was born in Berlin, Germany, into a 'wealthy run-of-the-mill assimilated Jewish family' (Arendt 1973: 33; see Jay 1973: 199; Roberts 1982; Wolin 1994). He was raised in a well-off quarter of the city and came of age during the Weimar Republic years before eventually settling in Paris in the 1930s. The historical record is patchy but apparently he earned his living, supported a wife and family until his divorce in 1930 and a passion for book collecting, from a combination of journal and newspaper publications, a stipend from the Frankfurt School, and by other 'private means', most likely his father, who was an art dealer and antiquarian.

The longevity and extent of Walter Benjamin's posthumous fame and influence is in inverse proportion to the relative brevity of his life, and the even shorter time-span of his academic and publishing output (Roberts 1982: 1–3). His publishing career spanned but a decade (Arendt 1973: 7). His early academic record was chequered, to say the least, in that a mixture of 'bungling and bad luck' (Arendt 1973: 14) dogged the reception of his work effectively preventing him from establishing a university career (Jay 1973: 203–204).[6] His two main pieces of scholarly research were published in 1920 and 1928, both of which were famously misunderstood at the time. It was only fifteen years after his death, with the publication of his collected work through the joint effort of Theodor Adorno and others, that his influence began to spread (Arendt 1973: 8; see Ferris 1996; Weigel 1997; Benjamin and Osborne 2000; Leslie 2000).

Perhaps the best known biographical detail of Benjamin's life is how it ended, with suicide at the age of 48 on the Franco-Spanish border in September of 1940. He was uneasily awaiting a visa that would allow him to emigrate to the United States of America, after fleeing to France from Nazi persecution. In Arendt's account, 'the immediate occasion for Benjamin's suicide was an uncommon stroke of bad luck' (Arendt 1973: 23–24; see Jay 1973: 197–198). Apparently, he mistakenly believed that he would not be able to obtain the necessary papers after being stopped at the Spanish border. Expecting to be sent back to Nazi Germany, he chose to kill himself instead. The historical and intellectual resonances of this personal choice have not gone unacknowledged by later commentators (Arendt 1973: 7; Jay 1973: 198).

A crucial aspect to Walter Benjamin's intellectual legacy is his role as co-founder of 'Critical Theory', the body of Marxist and Freudian influenced theory and research based at the University of Frankfurt (Jay 1973; see Devetak 1996; Best and Kellner 1991). His close – albeit stormy – intellectual relationship with Adorno and Horkheimer, the doyens of the Frankfurt School, is an important theme in the literature (Best and Kellner 1991: 219, 223; Jay 1973: 198–199). Benjamin, who 'was no-one's disciple' (Arendt 1973: 14; Leslie 2000: viii), was 'probably the most peculiar Marxist ever produced by this movement, which God knows had its full share of oddities' (Arendt 1973: 16–17); was involved in the European Communist movement – he visited the Soviet Union – and Zionist activism at the same time (Arendt 1973: 39; Jay 1973: 200–201); dreamt of publishing a work made up entirely of quotations in a pre-postmodern age; contributed to aesthetic and architectural theory and philosophy of history; was an accomplished translator; wrote (famously) about Goethe, Proust, Baudelaire and Kafka, book collecting, wandering about the city, and technological change (the focus of this chapter). This eclecticism is reflected in the vast quantity of secondary literature on his life and work. The main thing to remember for the interested reader is that caveats and arguments – about ideological affiliation, methodology, political applicability – abound when it comes to this thinker (see Wolin 1994: xxi; Leslie 2000: viii; Roberts 1982: 23).[7]

Historical and intellectual issues

> Benjamin . . . wants to engage in the world as he finds it . . . Materialism – in Benjamin's sense – assumes an interaction between people and world. Humans work upon physical things and materialism questions the ways in which they do this, and how this alters their thoroughly historical human nature.
>
> (Leslie 2000: ix)

Getting to grips with Benjamin does require some understanding, nonetheless, of the complex intellectual and historical currents in which he operated. These include the institutions and organisations he wrote for and who paid his wages; the various (strong-minded) prominent intellectuals who were his friends, mentors and admirers; the vicissitudes of (western European) Marxist theory and politics during the early twentieth century and since; the political economic repercussions of the inter-war years in which he lived and died (Roberts 1982: 3–5, 76 passim).

Let us return briefly to the oft-referred fraught relationship between Benjamin and the other founders of Critical Theory, who not only financially supported him from the mid-1930s but also edited and published his work (Roberts 1982: 70; Leslie 2000: 130–131; Wolin 1994: 163 passim; Jay 1973: 210–211). This relationship is not without interest for the history of ideas in general, let alone the legacy of any major school of thought or the intellectual development and significance of a thinker in particular (Roberts 1982: 73–74). The Artwork Essay's progression through various drafts for its publication in German, French and then English illustrates these tensions – philosophical, political and, no doubt, personal – between Benjamin and Adorno and, to a lesser extent, Benjamin and Horkheimer. Both criticised Benjamin for not being 'dialectical' enough, among other things (Jay 1973: 197 passim; Roberts 1982: 66 passim; Best and Kellner

1991: 217 passim). With respect to this essay's use of 'crude' historical materialist terminology (Roberts 1982: 153 passim) and the political exigencies of publishing in the USA (Jay 1973: 205), the sticking point for his two colleagues (who were at once fans as well as mentors) was that Benjamin was being way too optimistic about the revolutionary potential of new technologies to affect the structural power of capital. Moreover he was seen to be way too enamoured with popular culture and so too 'uncritical' about the political savvy of the 'masses', either under the sway of Fascist propaganda or as consumers of the capitalist 'culture industries'.[8] This difference of opinion was deeply felt for what Adorno and Horkheimer

> now feared was that mass art had a new political function diametri-
> cally opposed to its traditionally 'negative' one; art in the age of
> mechanical reproduction served to reconcile the mass audience to the
> status quo. Here, Benjamin disagreed . . . he paradoxically held out
> hope for the progressive potential of politicised, collectivised art.
>
> (Jay 1973: 211)

These sentiments are still echoed in many analyses of politics and technology from successors of the Frankfurt School and beyond (see Best and Kellner 1991; Jay 1973).

Which brings us to a more contemporary aspect to the historical and intellectual issues at stake when revisiting the Artwork Essay from a 'critical' political perspective (Leonard 1990). Given the interest in Benjamin from all manner of theoretical-methodological takes and political affiliations, the following tip for reading his work in the light of contemporary developments in ICTs is a good one:

> For those who seek to follow in Benjamin's footsteps run the risk
> of becoming mesmerised by the aura of his life and thought. Before
> they can be appropriated, his ideas . . . must be unflinchingly brought
> into contact with other intellectual traditions, as well as new histori-
> cal circumstances . . . The greatest disservice one could do to his
> theoretical initiatives would be to accord them the status of received
> wisdom, to assimilate them uncritically or wholesale. His mode of
> thinking, both alluring and allusive, invites commentary and exegesis,
> which should not be confused with adulation.
>
> (Wolin 1994: xxi; see also Leslie 2000)

It is as an important theoretical initiative that the Artwork essay is being treated here.

Initial parameters

To recall: the period in which Benjamin was writing was the lead-up to the Second World War; a period of 'mass movements' (Benjamin 1973: 215) in which Fascism gained a firm foothold in western Europe. On the other side of the Atlantic, mechanised manufacturing assembly lines, now known as the Fordist mode of production (see Harvey 1989), were hustling in the consumer society. At the same time, silent movies were giving way to talking pictures. Along with the popularity of film came other forms of 'mass media' such as glossy magazines, newsreels (an important vehicle for wartime propaganda), and the 'great historical films' of the Soviet Union (Benjamin 1973: 215). Benjamin, taking his cue from the Frankfurt School's suspicion of the belief that 'a technologically advanced society automatically embodied freedom and progress' (Best and Kellner 1991: 219), combined this with a Brechtian understanding of art – and theatre – as a politicised and empowering form of expression for the common person. He then looked at how the cult of 'High Art' and its high priests' control over access were being shaken up by techniques of mechanical reproduction. But these subversive tendencies had to be defended against their appropriation by Fascist, commercial and cultural elitist agendas.

The first parameter, then, is that Benjamin does not treat 'art' and 'technology' as mutually exclusive domains; of creative endeavour on the one hand and 'nuts 'n' bolts' (bits 'n' bytes) on the other. Both have material characteristics, aesthetic properties, and 'social significance' (see Leslie 2000: xii). Nonetheless, new techniques of mechanical reproduction do have an impact on creative processes. In doing so, they have the potential to challenge the political and cultural status quo.

Looking back from an era of 'global capital' and its concomitant gender–power relations (McChesney *et al.* 1998; Harvey 1989; Haraway 1990), the political and cultural lexicon of the Artwork Essay is particularly resonant. Benjamin was acutely aware of the political and social agenda of Fascism in Europe of the 1930s. He was also well aware of the new forms of economic and social exclusion developing with the rise of the consumer society, manufacturing assembly lines and increasing commodity fetishism in non-Fascist societies. Which bring us to the second thing to notice in Benjamin's approach. These new technical attributes have the potential to enable empowering change in societies marked by economic and social divisions. They can also disempower, which is why examining the relationship between the arts and technological change means raising political as well as cultural questions.

Another parameter for this reading is that Benjamin makes no secret of his admiration for the early political and technological successes of the Soviet Union. The (assumed) benefit of hindsight after the Stalinist era and the demise of the Communist Bloc in the late 1980s notwithstanding, this political stance does not diminish the power of his argument. By carefully analysing the mechanically reproducible pictorial and cinematic depictions of the natural and social worlds, Benjamin radically questions assumptions about 'authentic' art *and* politics. He argues that new repro-ductive techniques can – indeed they should – emancipate art from the stifling reign of the style police, on the one hand, and that their creative potential be used to alleviate the alienated labour relations underpinning capitalist economies on the other (Benjamin 1973: 218). Relative to his contemporaries' dismissive attitude to the aesthetic and/or social value of the new 'culture industries', this is a relatively up-beat approach to techno-logical change and its relationship to political economic and sociocultural change.

A final parameter pertains to the relationship between this essay, Benjamin's work as a whole and Benjamin-related literature. This piece can be read in its own right, and, furthermore, one need not be a Benjamin buff to be able to engage with what he has to say.[9] But it is a complex argument with many layers. This essay is much more than a nostalgic farewell to the 'aura of the [unique] work of art' (Benjamin 1973: 215) withering away before the 'making [of] many reproductions'. But neither is it an unequivocal celebration of the 'tremendous shattering of tradition' that is encapsulated in the 'destructive, cathartic aspect' (Benjamin 1973: 215) of popular film (or the Internet for that matter). On both counts, the essay is more than a quaint example of early technological determinism or naive political idealism. As Benjamin acerbically notes, changes in the arts and technology are not impervious to capitalist research and develop-ment agendas and totalitarian political ideologies. But neither are they untouched by their reception and appropriation by – and for – empowering ordinary people.

So let us now look at what Benjamin was actually saying in the English version that finally emerged out of the Artwork Essay 'débâcle' (Leslie 2000: 131).

Reading the Artwork Essay for digital days

This 'extremely concise and opaque' (Roberts 1982: 3) piece of writing looks at the social and political implications of the impact of 'mechanical reproduction' on 'the artistic processes' in both the so-called high and low cultural realms (see Wolin 1994: 208) under conditions of rapid techno-logical change. Or, more precisely, 'the nature of the repercussions that these two different manifestations – the reproduction of works of art and the art of the film – have had on art in its traditional form' (Benjamin 1973: 214). Benjamin argues that the 'age of mechanical reproduction' heralds significant changes in how industrialised societies perceive, experience, and then reproduce the world around them. Benjamin is interested in the socio-political ramifications of these changes in terms of how they radically alter the relationship between traditional art (and its philo-sophical, economic and moral underpinnings) and society at large. He argues that new technical capabilities (to re-present and distribute images), and the ways they challenge received aesthetic and cultural wisdoms, have 'revolutionary' potential.

Right at the outset, Benjamin thumbs his nose at both orthodox Marxist causal explanations and cultural elitism (see Roberts 1982: 153 passim; Wolin 1994: 183 passim). In a claim that is reiterated in the last lines of the essay (Benjamin 1973: 235), he contends that the 'develop-mental tendencies of art under present conditions of production [are a] dialectic [that] is no less noticeable in the superstructure than in the economy' (Benjamin 1973: 212). Benjamin plots out how the *material* attributes of film and photography challenge traditional artforms. Analysis, diagnosis, and prognosis are all woven into the fifteen theses that comprise Benjamin's journey 'back to the future' of contemporary techno-political change (see Arendt 1973).

His theoretical concern in examining the impact of these new tech-niques of mechanical reproduction on 'traditional' art is whether or not in the 'age of mechanical reproduction . . . the very invention of photography [and film has] not transformed the entire nature of art' (Benjamin 1973: 220). His answer is 'yes, it has'. While the prototypes for these latest reproductive techniques can be traced back into antiquity and then with the advent of lithography, they carry their own radical implications for contemporary and future generations. Given the moribund and esoteric condition of Art and Theatre at the time (Benjamin 1973: 214–215) this transformation need not be a bad thing for it enables more egalitarian and empowering forms of cultural and artistic expression.

Histories

The substantive analysis begins with a brief historical introduction into how 'in principle, a work of art has always been reproducible' (Benjamin 1973: 212). At the same time the subject matter shifted from deities and abstract depictions of beauty to ordinary people (family portraits) and their everyday life (at work, in the street). Photography's popular success set a new standard in 'pictorial reproduction' that was achieved by the camera rather than by the artist's hand. Moving pictures – film – took this freeing-up of the 'most important artistic functions' (1973: 213) much further. By the turn of the twentieth century, it had 'not only permitted it to reproduce all transmitted works of art and thus to cause the most profound change in their impact on the public; it also had captured a place of its own among the artistic processes' (1973: 213–214)

This new public dimension and the relationship between any work of art and its audience is a crucial element to these changes (see Wolin 1994: 186–187). The next few sections go on to unbundle his claim further.

Aura, perception, tradition

This new reproducibility – of a sacred image, person's face, landscape – makes a 'unique' object into one of many and then allows it to be made more readily available to many more people anywhere and at any time. For technical (non-manual) reproducibility can put 'the original in situations which would be out of reach for the original itself' (Benjamin 1973: 214). This shift 'represents something new' (1973: 212) that heralds the collapse of long-held assumptions about what constitutes uniqueness. Hierarchical divisions, between 'high' and 'low' culture, art-lovers and the general public (the 'masses') are turned upside-down. This has a profound effect on not only the business (in every sense of the term) of art(istic) production and reproduction, transmission and reception but also on the very nature of 'art'. This upsets the relationship between the authority vested in 'authentic' art and by way of concomitant assumptions about 'tradition' and 'historical testimony (1973: 215).

The upshot is 'that which withers away in the age of mechanical repro-duction is the aura of the work of art' (Benjamin 1973: 215; see Gasché 2000: 184; Jay 1973: 210).[10] That which was held to be sacred, is no more. What makes these new reproductive techniques socially significant is their ability to detach

> the reproduced object from the domain of tradition. By making many reproductions it substitutes a plurality of copies for a single existence. And in permitting the reproduction to meet the beholder or listener in his [sic] own particular situation, it reactivates the thing reproduced.
>
> (Benjamin 1973: 215)

For Benjamin, the 'most powerful agent' for this breakaway is film. The cathartic aspect to film-watching is that which makes it so particular, so fascinating and so politically potent. This is a double-edged sword in that this same quality is exploited by Fascist film propaganda. These cultural and political tensions are laid out in no uncertain terms in these opening sections.

Benjamin goes on to examine the relationship between 'sense perception' and 'social transformations' (Benjamin 1973: 216). He is less interested in working out the causal relationship between these than he is in how 'nature' and 'historical circumstances' both constitute the means and 'medium' through which the world is perceived. Changes in perception are also changes in spatial relations; or rather, assumptions and habits about what spatial relations mean (distance versus proximity for instance). The power of any aura resides in the physical and psychic gender–power relationships of physical distance and proximity vis-à-vis the object in question. Benjamin puts these changes partly down to the 'increasing significance of the masses in contemporary life' who want to get closer to things 'spatially and humanly' and so rejoice in having access 'to an object at very close range by way of its likeness' (1973: 216–217). That seeing the 'real thing' with the 'the unarmed eye' versus a reproduction differs is more to the point than whether one is superior to the other.[11]

In the fourth section, Benjamin looks at how *tradition* (see McCole 1993; Leslie 2000) is 'thoroughly alive and extremely changeable' (Benjamin 1973: 217). Following the spirit of the Frankfurt School (see Best and Kellner 1991: 215 passim), Benjamin argues that there is nothing inherent in the 'unique value of the "authentic" work of art' beyond its basis 'in the service of ritual' (Benjamin 1973: 217). The gatekeepers (secular and religious) of these rituals (cultural and political) to the 'first truly revolutionary means of reproduction, photography' (1973: 218) in the preceding century tried to maintain their power over what is/is not 'art'. Benjamin argues that such reactions simply underscore how 'for the first time in world history, mechanical reproduction emancipates the work of art from its parasitical dependence on ritual'. Break that relationship and the whole issue of what constitutes authenticity has to be reframed. Do that and art can become useful again by beginning to be 'based on another practice – politics' (1973: 218).

Cult and exhibition value: contemplation and interaction

If reproducibility can make traditional artworks more accessible and/or facilitate the popular success of new forms of (popular) cultural expression, then the artist and audience are brought into a much closer relationship than hitherto. Benjamin reflects on how this can be partly attributable to a gradual shift from the 'cult value' of artworks to their 'exhibition value' (where public showings supplant ritualistic ones). Together with reproducibility, this results in a different relationship between art and its public as well as 'a qualitative transformation of [art's] nature' (Benjamin 1973: 219). Shoring up against these forces by recourse to essentialist and ahistorical definitions of (great) art is a vain exercise as 'for better or for worse, technical forces have entered into the very heart of the process' (Wolin 1994: 190–191). And so have the responses of the 'great unwashed' – the general public.[12]

Photography exemplifies this entry of the arts into the public realm. Keeping the historical record close at hand, Benjamin is at pains to note that this is not an immediate or unchallenged shift seeing as early photography mimicked portrait painting. But as photography came into its own, with its own techniques and genres (picture captions, film story-boards, people-less landscapes and street-scenes), so did the need for a 'specific kind of approach' (Benjamin 1973: 220) to understanding the different quality of 'meaning-making' and the 'hidden political significance' of this new interactive dynamic. Film and photography's ability to activate a relatively shared interactivity (taking a photo, looking at photo albums, following and reacting to the plot) is quite different from that of 'free-floating contemplation' of an artwork in a hallowed place such as a museum, gallery, or temple (1973: 220). Techniques like those in photography and film have changed the nature of art and thereby the questions that need to be asked. Benjamin has no time for 'reactionary' art critics and theoreticians who turn up their noses at emergent artistic forms. But neither does he have time for those who in 'their desire to class the film among the "arts" . . . read ritual elements into it' (1973: 221). For Benjamin, the real issue is how art needs to be re-empowered and more so under the pressure of unequal social relations of capitalist accumulation.

The camera as performer and mediator

In the next four theses, Benjamin looks more closely at the actual techniques of photography and film-making vis-à-vis painting and theatre in order to underscore his point. Like new(er) ICTs, the former in their time affected processes of hand–eye coordination and the cognitive, and psycho-emotional processes by which (extant or new microscopic, telescopic or 'virtual') worlds were perceived and (re)presented.

The eighth and shortest section looks at the difference between acting on the stage and acting in front of the camera. The presence of the movie camera has several consequences to how the actor and audience interact to the situation, and how the each responds to the other. Unlike theatre, film performance is asynchronous, disembodied and then reconstituted by technical intervention; the 'position of the camera' (Benjamin 1973: 222) and the techniques of post-production editing (embedded techniques now taken for granted by later film-going and television-watching generations). The main point here, though, is the mediation of the relationship between actor/performer and spectator by the camera. Given that film meshes static pictorial reproduction (still applicable in photography) with dynamic ones (the emergence of nascent multi-media in effect), it directly challenges time-honoured notions of intact presence in artistic (re)production. New photographic/cinematic techniques mean both the performance and the actor can be recast at a later date (1973: 224). The upshot when meeting the final product (the film) is that the 'audience's identification with the actor is really an identification with the camera . . . [which] is not the approach to which cult values may be exposed' (1973: 224). And second, film 'strikingly shows that art has left the realm of the "beautiful semblance" which, so far, had been taken to be the only sphere where art could thrive' (1973: 224).

Empowerment or alienation?

Benjamin does not deny feelings of 'strangeness', 'estrangement felt before one's own image in the mirror', or 'new anxiety' for performer or spectator (Benjamin 1973: 224). But this blurring of the hierarchical distinction between 'author and public' (1973: 225) has not come out of the blue. It can be traced back to the advent of the printing press and popular journalism. However, here Benjamin reminds the reader of new conditions of 'capitalistic exploitation' and its role in the newer 'cult of the movie star' (1973: 226, 224). The 'shrivelling of the aura' is simply replaced by the 'phoney spell of a commodity'. Film alone, and certainly under conditions of capital accumulation where 'movie-makers' capital sets the fashion', does not a political/cultural revolution make. Other, more substantive changes in the unequal power relations of ownership and control have to occur for these technologies to enable new opportunities for disenfranchised groups (1973: 226).[13]

These broader conditions notwithstanding, cinematic (representations of) realities are qualitatively different by virtue of being technically mediated. The greater scale of intervention – agency – in the creative process is also different from that of the painter, or the surgeon and magician for that matter. Here, Benjamin is interested in comparing different sorts of interventions into different levels of 'reality' and perceptions thereof (Benjamin 1973: 226). Film actually ends up by creating an 'equipment-free aspect of reality'. Even as the authenticity (artistic merit) of a painter's interventions become assumed over time, the irony is that while the

> painter maintains in his work a natural distance from reality, the cameraman penetrates deeply into its web. There is a tremendous difference between the pictures they obtain. That of the painter is a total one, that of the cameraman consists of multiple fragments which are assembled under a new law. Thus . . . the representation of reality by the film is incomparably more significant than that of the painter, since it offers, precisely because of the thoroughgoing permeation of reality with mechanical equipment, an aspect of reality which is free of all equipment. And that is what one is entitled to ask from a work of art.
>
> (Benjamin 1973: 227)

This insight speaks directly to contemporary arguments about the aesthetic value of 'virtual' or computer arts, and concomitant reflections on 'virtual' versus 'real' creative activities.

Conscious enjoyment and unconscious optics

Benjamin goes on to argue that not only does the 'mechanical reproduction of art [change] the reaction of the masses toward art' but also this reaction simply underscores the sharp 'distinction between criticism and enjoyment by the public' (Benjamin 1973: 227). Benjamin sees the latter as an example of a 'progressive reaction'. Furthermore, this response is 'characterised by the direct, intimate fusion of visual and emotional enjoyment with the orientation of the expert. Such fusion is of great social significance' (1973: 227). Art and art criticism (destined for the few by the few) has become far removed from 'the critical and receptive attitudes of the public' (1973: 227). Instead, new art forms should be judged according to their 'social significance'; their ability to traverse the 'graduated and hierarchised mediation' that characterises the history and public life of traditional art (1973: 228). Popularity is not necessarily in inverse proportion to aesthetic worth in this regard. For Benjamin, the general public are not as stupid and undiscerning as they are made out to be by self-appointed style gurus. And the latter often have too much invested in the status quo to be able to understand that the rules have been changing.

These are complex processes, needless to say, which is why Benjamin draws upon psychoanalytic theories in order to underscore how the 'entire spectrum of optical, and now acoustical, perception' offered by film allows deeper, 'more precise statements' of everyday life and behaviour (Benjamin 1973: 229). Given the complex relationship between human consciousness (where the unconscious operates even when it is not immediately apparent or recognised) and social reality (which is not reducible to only observable phenomena), Benjamin celebrates film's ability to open up 'our taverns, and our metropolitan streets, our offices and furnished rooms, our railroad stations and our factories [that] appeared to have us locked up hopelessly [so allowing people to] calmly and adventurously go travelling' (1973: 224). No fear of 'virtual' realities here as Benjamin argues that the empowering potential of mechanical (and digital) reproduction lies in this revealing of 'entirely new structural formations of the subject'; this intervention into 'unconsciously penetrated space'. The 'camera introduces us to unconscious optics as does psychoanalysis to unconscious impulses' (Benjamin 1973: 230).

Perception and action: on and off-screen

As always with Benjamin, these liberating tendencies cannot be divorced from broader political economic and cultural forces. Benjamin draws the essay to a close by recalling how such shifts, revelations and subjugations can only be seen in 'critical epochs in which a certain art form aspires to effect which could be fully obtained only with a changed technical standard, that is to say, in a new art form' (Benjamin 1973: 230). The 1930s were one such 'critical epoch' for the emergence of changed technical standards (namely art forms). In the last two theses, Benjamin acknowledges, in a brief comparison to the Dadaist art movement, that not all film is progressive (1973: 231) just as not all existing art is reactionary.[14] And there lies the rub for many critics of his optimistic take on the popular arts (see Wolin 1994: 197 passim). For what is at stake are the boundaries between material attributes, gender–power relations, and taste. Here, Benjamin is more interested in how the 'transmutations' in art and technology of the time constitute a qualitative shift towards a 'new mode of participation' (Benjamin 1973: 232) than he is in deciding between 'good' or 'bad' art. Popularity also needs to be taken seriously in that this 'greatly increased mass of participants has produced a change in the mode of participation'. For commentators to assume that this is a superficial development is 'at bottom the same ancient lament that the masses seek distraction whereas art demands concentration from the spectator' (1973: 232). This is not only a specious separation but also an elitist assumption of what constitutes 'art' in the first place and its facilitating role in political struggle.[15]

The final thesis argues for more attention to be paid to how the arts can engage the other senses and cognitive processes (Benjamin 1973: 233); that is if they are to have any use at all apart from their commercial viability. Film, for Benjamin, has potential because of its ability to make traditional

> cult value recede into the background not only by putting the public in the position of the critic, but also by the fact that at the movies this position requires no attention. The public is an examiner, but an absent-minded one.
>
> (Benjamin 1973: 234)

The Epilogue

The Epilogue returns to the initial grand themes of the Prologue. Techniques of mechanical reproduction are appropriated by the commercial film industry, the art establishment and oppressive political agendas which threaten the chance to reinvigorate and liberate 'artistic processes' from the strictures of 'art in its traditional form' (Benjamin 1973: 214). Benjamin lays the blame for this at the feet of capitalism's processes of commercialisation with its 'growing proletarianisation of modern man' (1973: 234), and the Fascist war machine's use of mass mobilisation (through the propaganda film and hi-tech weaponry). In the first instance, the

> capitalistic exploitation of the film denies consideration to modern man's [sic] legitimate claim to being reproduced. Under these circumstances the film industry is trying hard to spur the interest of the masses through illusion-promoting spectacles and dubious speculations.
>
> (Benjamin 1973: 226)

In the second, the return to 'tradition' and the celebration of 'authenticity' by Fascist regimes in Italy and Germany

> organize the newly created proletarian masses without affecting the property structure which the masses strive to eliminate. Fascism sees its salvation in giving these masses not their right, but instead a chance to express themselves. The masses have a right to change property relations; Fascism seeks to give them an expression while preserving property . . . The violation of the masses . . . has its counterpart in the violation of an apparatus which is pressed into the reproduction of ritual values.
>
> (Benjamin 1973: 234)

For Benjamin, the sad irony is that the nascent changes in 'human sense perception' allowed for by these technical innovations and their creative potential (1973: 217, 218–219) have been sucked into serving the needs of war on the one hand and capitalist expansion on the other. Not only does war become 'the artistic gratification of a sense perception that has been changed by technology' (1973: 235) but also there is an all too painful

> discrepancy between the tremendous means of production and their inadequate utilization in the process of production . . . Instead of draining rivers, society directs a human stream into a bed of trenches; instead of dropping seeds from airplanes, it drops incendiary bombs over cities; and through gas warfare the aura is abolished in a new way.
>
> (Benjamin 1973: 235)

The Epilogue's seeming tone of despair only really makes sense in the light of how the preceding 'deconstruction' of film and photography – as techniques and aesthetics – highlights new technologies' ability to drive a wedge between received wisdoms about art, technology, and politics. By virtue of their popularity, if nothing else, they can challenge for whom, and by whom, art (or politics) is made (1973: 244, note 21). But there is also a tension between these competing appropriations and uses of new technologies/art forms, political mobilisation and everyday perception and daily habits. The window of opportunity is but a small one.

Benjamin in dialogue with cyborgs

[F]or the first time in world history, mechanical reproduction emancipates the work of art from its parasitical dependence on ritual. To
an ever greater degree the work of art reproduced becomes the work
of art designed for reproducibility . . . [The] instant the criterion of
authenticity ceases to be applicable to artistic production, the total
function of art is reversed. Instead of being based on ritual, it begins
to be based on another practice – politics.

(Benjamin 1973: 218)

It is not just that science and technology are possible means of great
human satisfaction, as well as a matrix of complex dominations.
Cyborg imagery can suggest a way out of the maze of dualisms in
which we have explained our bodies and our tools to ourselves . . .
I would rather be a cyborg than a goddess.

(Haraway 1990: 223)

It would be tempting at this point to assign Benjamin's 1930s analysis
of the impact of relatively 'old' technologies on even older traditions in
art and culture to posterity; to underscore the many differences between
mechanical and digitalised technologies (see Spiller 2002). But as others
have already shown (Leslie 2000; Roberts 1982; Benjamin and Osborne
2000: ix–xiii), this not only does a disservice to the cogency of Benjamin's
thought but also assumes that contemporary debates about technological
– socio-cultural – political economic change have developed onwards
and upwards; encapsulated in the hi-tech, digitalised, societies of today.
Not necessarily. By the same token, however, neither is past scholarship
sacrosanct. In both cases, feminist and/or postmodernist critiques in recent
years have focused on how bodies of theory and research are just as
beholden to their own 'cult values', 'traditions' and categorical 'auras'
as are the arts (Harding 1998; Haraway 1992; Docherty 1993; Nicholson
1994). Walter Benjamin, like others (see chapters in this volume), needs
to be called upon without distorting important differences or similarities
in intellectual, cultural or political economic climates. To underscore this
delicate balance between historicity and new-ness (see Leslie 2000: x),
the rest of the chapter juxtaposes the Artwork Essay with a key work by
an important theorist of 'the age of digital reproduction'. Alive and well
and living in California, USA, Donna Haraway has made an important
contribution to thinking about the ramifications of ICTs for the relationship

between human-being-ness and machines.[16] She also writes – contro-
versially – from out of the Critical Marxist tradition of social theory and
cultural critique that incorporates both postmodern and feminist insights.[17]

Conversations

> The structural relations related to the social relations of science
> and technology evoke strong ambivalence . . . For excellent reasons,
> most Marxisms see domination best and have trouble understanding
> what can only look like false consciousness and people's complicity
> in their own domination in late capitalism. *It is crucial to remember*
> *that what is lost, perhaps especially from women's point of view, is*
> *often virulent forms of oppression, nostalgically naturalised in the*
> *face of current violation.* Ambivalence toward the disrupted unities
> mediated by hi-tech culture requires not sorting consciousness into
> [dichotomous] categories . . . but subtle understanding of emerging
> pleasures, experiences, and powers with serious potential for changing
> the rules of the game.
>
> (Haraway 1990: 214–215, emphasis added)

This statement, echoing Benjamin in more ways than one, is from
Donna Haraway's highly acclaimed *Manifesto for Cyborgs* (1990) first
published in 1985 (Spiller 2002: 108–109). At that time, fifty years after
Benjamin published the Artwork Essay, high capacity and polluting
assembly lines of heavy manufacturing were making way for 'clean' just-
in-time production lines of the 'knowledge economy' and microelectronic
manufacturing. Computerised techniques had taken hold in the printing
and film industries and the 'postmodernist' critique of western enlightenment thought had begun in earnest. Broad political mobilisation from the
Left was also sagging under the effects of de-unionisation as neo-liberal
macroeconomic orthodoxies sacrificed employment to utopian visions
of monetary stability. Somewhere at the epicentre of all these complex
changes (whose constitution and ramifications are by no means decided
upon as yet) was the emergence of digitally integrated information and
communication technologies. For Haraway these have come to define
'an emerging system of world order analogous in its novelty and scope to
that created by industrial capitalism: we are living through a movement
from an organic, industrial society to a polymorphous, information system
. . . from the comfortable old hierarchical dominations to the scary new
networks . . . called the informatics of domination' (Haraway 1990: 203).

Having said this, Haraway then goes on to greet the 'postmodern, non-naturalist' mood of the day (1990: 192) in an 'argument for pleasure in the confusion of boundaries and for responsibility in their construction' (1990: 191) as opposed to an entrenchment into reified (Marxist and feminist) categories and political dogma. With an explicit attack on longstanding 'leaky' separations between 'human and animal', 'animal-human (organism) machine', and 'physical and non-physical' (1980: 183, 195), Haraway – like Benjamin – seeks to regain a political prerogative by reclaiming metaphorical and rhetorical territory dominated by the 'New Right' – and the 'old Left' – in a time characterised by 'the extent and importance of rearrangements in worldwide social relations tied to science and technology' (Haraway 1990: 203). Remembering that this was the era of Reaganomics in the USA and Thatcherite economic and cultural policies in Britain (Hall 1996), Haraway's political economic critique is focused on the appropriation of ICTs for neoliberal economic and/or conservative cultural agendas. Most particularly, she is concerned with how these affect those women (especially from ethnic minorities) as a newly exploitable and vulnerable labour force (Haraway 1990: 207, passim). Her approach is squarely placed within Marxist/feminist critiques of capitalist/patriarchal societies in a deliberate 'effort to build an ironic myth faithful to feminism, socialism, and materialism' (1990: 190) that uses irony as a 'rhetorical strategy and a political method . . . At the centre of my ironic faith, my blasphemy, is the image of the cyborg' (1990: 190–191).

While Benjamin celebrates the new capacities of photography and film to subvert and open up sense perception and everyday experience for ordinary people, Haraway urges a comparable embracing of 'fabricated hybrids of machine and organism' (Haraway 1990: 191) in order to claim these for political mobilisation. With her critical eye on Marxist and/or feminist orthodoxies that would spurn all recourse to 'postmodern' subject matter and vocabulary, Haraway is also interested in the 'social signif-icance' (Benjamin 1973) and political ramifications of this generation of technical (re)producibility. Instead of their immediate facilitation of 'pictorial reproduction', Haraway looks at how ICTs radically reconstitute the biological body, and women's bodies in particular, both as potentially empowering 'cybernetic' organisms and as exploitable workers in a new mode of production (Haraway 1990: 212 passim). While Benjamin exam-ines the repercussions of the withering away of the aura of an artwork on the basis that the former is a (powerful) social and historical construction, Haraway embraces the metaphor of the cyborg as an 'imaginative resource suggesting some very fruitful couplings' (1990: 191). Borrowed from science fiction mainly, the cyborg is 'simultaneously animal and machine, who populate worlds ambiguously natural and crafted' (1990: 191). Haraway posits it as a utopian 'creature in a postgender world' (1990: 192) that can operate as a metaphor for a 'myth of political identity' (1990: 215) that empowers women – and men – under current conditions of 'racist, male-dominated capitalism' and its particular 'social relations of science and technology' (1990: 191, 217).

Haraway's feminist rhetoric and explicit postmodern take on the social and historical contestability of dominant western technological culture and gender–power relations is not as discordant with the Artwork Essay as it might first appear. Both Benjamin and Haraway use the more fluid essay genre, resonant metaphors and left-wing political rhetoric to problematise sacrosanct assumptions about the integrity of art and culture, technology and politics. Both entertain new forms of cultural expression and ontologies in order to productively respond to the times as opposed to (over)react. And even as both celebrate newness they also resist being hopelessly optimistic about mechanical reproducibility or cybernetic crossovers *per se*. For both are examining emergent technologies in terms of by whom – and for whom – they are being appropriated. Both broach the gender–power politics of control and ownership, and access to new cultural forms and political possibilities.

Implications for theory and research

Let me sum up by touching on some of the implications of this brief encounter between two 'blasphemous' thinkers (Haraway 1990: 190) for theory and research. First, Benjamin's unabashed fascination with the creative and political potential of mechanical reproduction does not preclude him from critically analysing its impact on the physical senses, the individual and social (un)conscious. His political angle also precludes contemporary concerns about the ramifications of hi-tech forms of warfare, genetic engineering and intrusive surveillance techniques, to which Haraway's Cyborg Manifesto directly speaks. Whether these are aided and abetted by the spread of World Wide Web-based communications and the boom-and-bust of the Internet gold-rush or not, the need to ask new questions persists. Unlike the 1930s though, by the turn of the twenty-first century, the composition of political party affiliations had shifted from clearly signposted ideologies of left-wing or right-wing. Even as the technologies in question, the cultural references, and the 'change agents' in these two essays are not identical, both authors treat all these as sites of political and cultural contestation rather than foregone conclusions.

Second, I have read the Artwork Essay in terms of how Benjamin is 'prequeling' approaches, such as Haraway's, that stress the socially constructed nature of techno-cultural change (Haraway 1997b; Harding 1998). He is interested in just how delicately balanced these processes are for political struggles and artistic expression and all the boundaries and crossovers that there are between the technological, cultural and political economic realms. For Benjamin, war epitomised the *immaturity* and savagery of industrialised societies in this respect (Benjamin 1973: 235). Haraway is also deeply concerned about technological misappropriation on the one hand and retreat into reactionary myth-making on the other. Her point is that 'we are not dealing with a technological determinism but with a historical system depending upon structured relations among people' (Haraway 1990: 207). Both these thinkers' political insights are being borne out by much contemporary research and development into ICTs, which is largely funded by huge commercial and political strategic interests. These operate to produce comparable 'mechanical [digital] equipment' that feeds off 'a sense perception that has been changed by technology' (Benjamin 1973: 235) and then justifies the products (hi-tech biological weapons, anti-missile defence systems, punitive surveillance techniques) as either responses to 'consumer demand' or threats to national security. As in the 1930s, and even though the novel techniques of mech-anical reproduction Benjamin examines have become old-hat these days, a comparable set of struggles for ownership and control of new(er) techniques of reproduction – and destruction – are emerging in the age of digital reproduction. Rather than shying away, Benjamin for his part embraces new art-forms, while Haraway embraces new life-forms as ways of turning the tables.

A third theme relates to achieving nuanced, non-universalising analyses of ICTs from within a critique of capitalism and commitment to political and social empowerment that is beholden to Critical Theory (Burchill and Linklater 1999; McChesney *et al.* 1998; Roberts 1982). Problem is, many critics from this tradition lean towards a more jaundiced view of the shifts, particularly in terms of how they relate to the rise of postmodernism on the one hand and neo-liberal economics in the 1980s on the other (see Best and Kellner 1991; Harvey 1989). Space does not allow for a full explora-tion of this point. But suffice it to say this vein of critique sees digital reproduction's impact on 'art' and 'culture' as a negative force. According to this standpoint, one that takes its cue from Adorno's critique of the mass culture industry rather than Benjamin's (Best and Kellner 1991: 181 passim, 215 passim), there is more being lost than gained. Needless to say, I am referring to a range of sophisticated and pertinent analyses.

Nevertheless, the gist of these critiques is that aesthetic sensibilities are being corroded by ICTs 'unholy' alliance with late capitalism (see Harvey 1989; Huyssen 1990; Jameson 1984; McChesney *et al.* 1998).

In this respect, the ultimate aesthetic judgement of this perspective is that twentieth-century architectural developments, science fiction films and literature, interactive computer games, graffiti art, rap/hip-hop or electronic musical forms are all symptoms of this corrosive shift. The concomitant predominance of a high degree of simulation, multi-imaging, conscious copying or 'sampling' that is characteristic of digital repro-duction underscore how popular forms of art in the digital/Internet/ information age continue to challenge the authority of the 'unique' work or art as the ultimate form of artistic expression. Moreover, these latest challenges to traditional art – and its aura – are assumed to be inherently damaging for organised politics, if not the individual (Dufour 2001). Their uptake by disadvantaged groups and/or popularity is then all too easily dismissed as more evidence of their anti-aesthetic or politically suspect qualities. Both Benjamin and Haraway's interventions address some of the deeper aesthetic and ontological assumptions operating in these attitudes.[18]

Fourth, both these pieces were written as think-pieces to spur discussion on fundamental issues. Hence the many ways in which both the Artwork Essay and the Cyborg Manifesto have been interpreted and applied. As Benjamin argues, the problem is that in reducing new art forms and techniques of (digital) reproduction to their circumstantial link with ('late') capitalism and/or fundamentalist politics, or reifying them by confining them to tried and true categories, their emancipatory dimensions are pushed underground. More to the point, in so doing, these new spaces and places for 'human sense perception' to go travelling, for everyday exploita-tive power relationships (based on hierarchies of class, hetero/sexuality, sex/gender, race/ethnicity, religion) to be readdressed and challenged through the 'cathartic' effect (among others) of new art-and-technology are filled in by far less tolerant forms of political mobilisation. The challenge to cultural and social closure (most particularly from the point of view of non-western forms of expression in the arts) that these newer techniques offer, and their potential to open up other spaces for mobil-isation and expression are either ignored or condemned to dichotomies such as the 'clash of civilisations' between Christianity and Islam, global homogeneity versus local heterogeneity and so on. Without conflating the early twentieth and twenty-first centuries, Benjamin's focus on the material, symbolic, social and political dimensions to art/culture, politics and technology as they interact is still powerful. It can converse with Haraway's distinctive approach because both thinkers are arguing against blind faith in mutually exclusive and moribund cultural and political categories and the privileges they confer (Haraway 1990; Peterson and Sisson Runyan 1999).

Reading Benjamin for digital days is not without its problems, nevertheless given the different political and philosophical hues the prefix *critical* has acquired nowadays. It now designates intense debates between Marxist and non-Marxist scholars at least since the late 1960s (Best and Kellner 1991). In addition there has been criticism of the armchair politics of many of these thinkers and the Frankfurt School especially (Leonard 1990; Best and Kellner 1991; Roberts 1982: 66; Nicholson 1990; Ferguson and Wicke 1994). What these debates point to is an underlying ontological and epistemological tension when analysing the interaction of humans and/or machines; one that preoccupies both the Artwork Essay and the Cyborg Manifesto. Namely, this sort of 'criticism, like resistance, cannot exist without being complicit with what it criticises and thereby resists' (Ferris 1996: 2). This goes hand in hand with finding, and creating, practicable exit strategies from the 'informatics of domination' (Haraway 1990) or the 'aestheticisation of politics' (Benjamin 1973). Here, feminist and postcolonial scholars' attention to the gender–ethnicity–sexuality dimensions of these processes resonate with Benjamin's focus on openings rather than on apparent closure and loss. They do so because both these generations of critical Marxian analyses focus on the 'particularities and the most microscopic details of everyday experience [in order to] . . . analyse the major social processes, the mediating institutions and structures, that help constitute particularities' (Best and Kellner 1991: 223).

Conclusion

Written at a technological and historical cusp, the Artwork Essay is an example of reflexive theorising that looks to techno-cultural change as a harbinger of hopeful possibility rather than lost opportunities. This chapter has reread the Artwork Essay, arguably one of the most alluded to and argued about works of the twentieth century (Wolin 1994: 184; Leslie 2000: 130 passim), for its continuing relevance to critical theorising on the immediate impacts, relevance and ramifications of new(er) ICTs in the political economic and socio-cultural spheres. This essay succeeds in this way because it engages with the tensions between developments in mechanical (and now digital) reproduction, changes in political economic power relations, art and culture, and the ambivalence of reception to the same. It offers a productive way of dealing with this intellectual and emotional ambivalence – then and now.

This chapter has also read Benjamin, alone and in conversation with Haraway, for his contribution to conceptualising technological change *as it is occurring* (Wolin 1994: 185). As Benjamin himself concurs, it is difficult keeping complex 'developmental tendencies of art under present conditions of production' (Benjamin 1973: 212) in mind without lurching between an uncritical celebration of all that is new or a nostalgic lament for the (often exclusionary) cultural and political traditions that are under threat. This reading highlights how Benjamin's political commitment intersects with Haraway's in terms of their recognition of the empowering dimensions to seemingly disempowering changes. Theory and research into ICTs, the information society and so on needs to take account of these radical takes on the political potential of major technological change from within the Critical/Marxist/feminist traditions (*sic*). Under the current conditions of commodified research and development into ICTs Benjamin and Haraway provide ways to achieve this delicate balance between the critical circumspection of 'engaged political critique' (Leslie 2000: vii) and a rigorous, non-positivist approach to socio-cultural-technological change (Roberts 1982: 157–158; see Wolin 1994: 164). And do so from the point of view of those groups who do not control their current trajectory.

Precisely because of the wealth of Benjamin studies (see Ferris 1996: 3; Leslie 2000), this chapter has concentrated on the political import of his analysis of the relationship between technological change, politics and the arts. It has done so in order to recall that the strength of critical thought, such as that of Benjamin and Haraway, is its ability to elucidate nuance at moments of significant social and/or technological change. Taking risks is part and parcel of these two exercises. Taking a critical stance that engages with the political and cultural concerns of the day and to do so with theoretical and empirical rigour entails not only a healthy circumspection towards the glitter of all that is new – let alone the beckoning of the familiar – but also an awareness of the potential for empowerment contained in the same. In other words, Benjamin's approach provides an important entry-point and substantive insight into the psychic-emotional and political economic tension between how new ways of (re)presenting and (re)producing the world technically can challenge the status quo. Art and cultural life are not divorced from either technological change or concomitant socio-economic divisions and political struggle. And how emotional and material realities are (re)presented and (re)produced are not necessarily emancipatory or alienating by virtue of being mechanically, or digitally, reproducible. It is more a question of whose realities, and reproductions thereof, hold sway in the interaction between the arts, politics, and the vagaries of everyday human sense perception. This can create new spaces for practically mobilising without assuming that the current (neo-liberal) capitalist agenda for ICTs is a fait accompli, the 'only alternative'.

Notes

1   These entail all computer-based technologies that process and store information, mediate telecommunications traffic, facilitate the media industries, and contemporary medical intervention (see Haraway 1990: 206; Spiller 2002).
2   It is a moot point as to where the dividing line is between these two realms. This hierarchy, let alone the very definition of 'culture' and/or 'art' is hotly disputed in media and communication studies and cultural studies (Hall 1996; Best and Kellner 1991; Huyssen 1990).
3   The date refers to the three versions at stake in any study of this work. The German title is *Das Kunstwerk im Zeitalter seiner technischen Reproduzierbarkeit* (1936). The English-language version of this title stresses *reproduction* rather than *reproducibility* (see Leslie 2000).
4   Allusions to the Artwork Essay abound. Jay (1973: 205 passim) and Buck-Morss (1977, 1992) deal with it from the point of view of the Frankfurt School's intellectual and philosophical peregrinations. Wolin (1994) also has a chapter discussing its role in – and articulation of – Benjamin's theoretical 'dispute' with Adorno (Wolin 1994: 183 passim). Roberts (1982: 157 passim) looks at it in the context of Benjamin's application of historical materialist method and conceptualisation of technology – *Technik*. See also Leslie (2000) for an in-depth study of Benjamin that focuses on countering 'the hyper-cyberbabble of the new millennialism' (2000: x). Her chapter on the essay (Leslie 2000: 130 passim) takes into account all three versions in order to look more closely at 'Benjamin's critical breakdown of *Technik* into component parts . . . It is on this basis that Benjamin grounds a strategy for a critical political practice that utililizes technology in a "truly revolutionary way"' (2000: 133). See also the lucid reading by Gasché (2000) in a chapter that links Benjamin to Kant. Weber (1996: 32 passim) looks at it in his comparison of Benjamin and Heidegger. McCole (1993: 5–8) focuses on the essay as Benjamin's position-statement on the social and political implications of the 'vertiginous, disorienting acceleration of the pace of social and technological change in the opening decades of the twentieth century' (McCole 1993: 1). The volume edited by Fischer (1996) has a number of short reflections on Benjamin that allude to this essay from the point of view of gender, ecology, and contemporary theatre and performance art. And the list goes on. See Leslie (2000) for a thorough and up-to-date bibliography.
5   See Haraway (1990, 1991, 1997a). There are also copious Donna Haraway-related websites as her work not only has many fans but also intersects with science fiction literature, a source she herself draws upon in later versions of the Cyborg Manifesto. Her work is best known, however, in the fields of science and technology studies and feminist theory (Haraway 1997a, 1997b). Haraway writes from within a rich body of feminist theory on science and technology that dates back to the 1970s at least (see Haraway 1990: 225–226; Harding 1986; Spiller 2002). To echo Neil Spiller on the Manifesto itself,

Haraway is 'one of the very few . . . [who] adequately expresses a mature political thesis and a thoughtful feminist view' of technological change (Spiller 2002: 109).

6 As one commentator astutely notes in the light of how often this aspect to Benjamin's rocky career path is stressed:

> pursuit of a career is not always compatible with the maintenance of a high intellectual independence. But that does not mean that the reverse – maintenance of intellectual independence without any regard for career – is itself of any value . . . Benjamin neither sought nor believed in isolation . . . And the story of his 'career' is the story of a struggle to exploit unsympathetic or refractory organisational structures for his own purposes.
>
> (Roberts 1982: 23–24)

7 Leslie (2000: 219–225) has a very good section on the history and impact of 'Benjamin Studies' on the reception of his work. See also Arendt (1973: 9–10).

8 This partly related to there being some disaffection with his friendship with the Marxist playwright Bertolt Brecht (Jay 1973: 201–203; Wolin 1994: 139 passim) as well as the specific nature of Adorno and Benjamin's working relationship (Roberts 192: 71–73; Wolin 1994: 163 passim; Jay 1973: 205 passim).

9 For those who want to read more, or dip into Benjamin as opposed to immerse themselves in his *Correspondence* or *Collected Works*, the edited collection *Illuminations* is a good place to start. Another is *One Way Street and Other Writings* (1979), which include his *Short History of Photography*.

10 The term, *aura*, is a complex one. Benjamin's own definition in a footnote will suffice for now:

> The definition of the aura as a 'unique phenomenon of a distance however close it may be' represents nothing but the formulation of the cult value of the work of art in categories of space and time perception. Distance is the opposite of closeness. The essentially distant object is the unapproachable one.
>
> (Benjamin 1973: 236, note 5)

11 One only has to think of the mystique surrounding seeing the 'real' Mona Lisa in the Louvre, recall debates about the artistic merit of Mona Lisa reproductions (whether as postcards, T-shirts or Andy Warhol's screen-prints), or note the huge financial stakes involved in the discovery of (and market in) forged and/or authentic 'masterworks' to get Benjamin's point here.

12 In the context of post/modernism, these debates have continued to rage through the advent of Pop Art, postmodernist architectural movements and now 'digital aesthetics' (Harvey 1989; Cubitt 1998; Huyssen 1990).

13 At the time of writing, Benjamin's focus was Europe and North America. In the twenty-first century these dynamics are usually understood under the rubric of 'globalisation' however construed. In both periods, broadly speaking, essentialist and fundamentalist political/religious/cultural discourses and axes for populist politics were, and are, at work.

14 This anti-art establishment movement of the early twentieth century, which borrowed elements from surrealism and futurism, was based on keeping the

public on their toes, if not outraging them by deliberately flaunting artistic conventions. Dadaists saw 'art' as much as 'happenings' as formal properties like photo-montage or phonetic poetry. Key figures were Marcel Duchamp and André Breton.

15 This is also at the heart of the aforementioned disagreements between Benjamin and Adorno (see Wolin 1994; Jay 1973; Best and Kellner 1991). Space prevents a closer look at these debates. Suffice it to say, they also speak directly to the 'currency' (*Aktualität*) of this essay (Wolin 1994: 205; Leslie 2000: ix) for comparable arguments about digital aesthetics, digital democracy and cyber-activism.

16 This section is a revised version of a more in-depth comparison of Benjamin and Haraway (Franklin 2003). Despite her explicit Marxist feminist affiliations, Haraway is often grouped with other cyberspace gurus like Marshall McLuhan and Nicholas Negroponte even though they all have very different premises and conclusions (see Spiller 2002). In any case, they have all published influential reflections on cybernetics/cyborgs, computer-mediated representations of reality and communicative practices, digital ontological implications of ICTs – the Internet more precisely – in one way or another. In so doing, they have contributed crucially to contemporary conceptual categories and frameworks in academic research and the popular imagination. They have also been received, and regurgitated, in all manner of ways since the publication of their key works. Moreover, juxtaposition was a conscious device in Benjamin's critical method (see Best and Kellner 1991: 227; Arendt 1973: 51–53).

17 Linda Nicholson (1990) does one of the best jobs in summing up and collating key examples of these complex interactions between (Marxist) feminism and postmodernism. Ferguson and Wicke (1994) is a more recent collection along the same lines. Best and Kellner (1991) look at the Frankfurt School, Marxism, and feminist theory under the impact of postmodernist critiques of Enlightenment thought and historical shifts from modern to postmodern societies. In short, both these entail philosophical moves that question 'the dominant goals and assumptions informing modern theories of society, history, politics, and the individual, while embracing a variety of new principles and emphases' (Best and Kellner 1991: x). While their treatment of feminism is cursory, their study is very helpful in navigating this difficult and deeply polemical terrain. Huyssen (1990) provides another angle on 'this thing called postmodernism' from the point of view of art and architecture which intersects with Benjamin's interests as well. Harvey (1989) is an influential condemnation of the same shifts in the cultural sphere.

18 As many studies tend to focus on Benjamin's philosophical and aesthetic language, Esther Leslie's focus on how his conceptualisation of technology as socially and historically constructed is an important addition to the literature. She argues that in Benjamin, the term *Technik*

> intimates a sense of both technology and technique. Benjamin seems to squeeze full meaning from this compact word. In signifying simultaneously technology and technique, *Technik* alludes to the material hardware, the means of production and the technical relations of production.
>
> (Leslie 2000: xii)

Her in-depth analysis of his work, and the Artwork Essay in particular, based on the way he employs this term on different levels intersects with 'social constructivist' views of technological and political change that also critique capitalism (Haraway 1997b; Harding 1998). Julian Roberts also pays attention to this relationship (Roberts 1982: 157, 186).

References

Arendt, H. (1973) 'Introduction: Walter Benjamin: 1892–1940' in W. Benjamin, *Illuminations*, London: Fontana.

Benjamin, A. and Osborne, P. (eds) (2000) *Walter Benjamin's Philosophy: Destruction and Experience*, London: Clinamen Press.

Benjamin, W. (1973) [1935–1939] 'The work of art in the age of mechanical reproduction' in W. Benjamin, *Illuminations*, London: Fontana.

Benjamin, W. (1979) *One Way Street and Other Writings*, London: New Left Books.

Benjamin, W. (1986) *Reflections*, New York: Shocken.

Best, S. and Kellner, D. (1991) *Postmodern Theory: Critical Investigations*, London: Macmillan Education.

Buck-Morss, S. (1977) *The Origin of Negative Dialectics: Theodor Adorno, Walter Benjamin and the Frankfurt Institute*, Brighton: Harvester Press.

Buck-Morss, S. (1992) 'Aesthetics and anaesthetics: Walter Benjamin's artwork essay reconsidered', *October* 62 (fall): 3–41.

Burchill, S. and Linklater, A. (eds) (1996) *Theories of International Relations*, London: Macmillan Press.

Cubitt, S. (1998) *Digital Aesthetics*, London: Sage.

Devetak, R. (1996) 'Critical theory' in S. Burchill and A. Linklater (eds) *Theories of International Relations*, London: Macmillan Press.

Docherty, T. (ed.) (1993) *Postmodernism: A Reader*, London: Harvester Wheatsheaf.

Dufour, D.-R. (2001) 'Les désarrois de l'individu-sujet', *Le Monde Diplomatique* (February): 16–17.

Ferguson, M. and Wicke, J. (1994) 'Introduction: feminism and postmodernism; or, the way we love now' in Margaret Ferguson and Jennifer Wicke (eds) *Feminism and Postmodernism*, Durham, NC: Duke University Press.

Ferris, D.S. (1996) 'Introduction: aura, resistance, and the event of history' in D.S. Ferris (ed.) *Walter Benjamin: Theoretical Questions*, Stanford, CA: Stanford University Press.

Fischer, G. (ed.) (1996) *With the Sharpened Axe of Reason: Approaches to Walter Benjamin*, Oxford: Berg.

Franklin, M. (2003) 'Reading Walter Benjamin and Donna Haraway in the age of digital reproduction', *Information, Communication and Society* 5(4) (in press).

Frisby, D. (1996) 'Walter Benjamin's prehistory of modernity as anticipation of postmodernity? Some methodological reflections' in G. Fischer (ed.) *With the Sharpened Axe of Reason: Approaches to Walter Benjamin*, Oxford: Berg.

Gasché, R. (2000) 'Objective diversions: on some Kantian themes in Benjamin's "The work of art in the age of mechanical reproduction"' in A. Benjamin and P. Osborne (eds) *Walter Benjamin's Philosophy: Destruction and Experience*, London: Clinamen Press.

Hall, S. (1996) 'New ethnicities' in *Stuart Hall: Critical Dialogues in Cultural Studies*, London: Routledge.

Haraway, D. (1990) [1985] 'A manifesto for cyborgs: science, technology, and socialist feminism in the 1980s' in L. Nicholson (ed.) *Feminism/Postmodernism*, New York: Routledge.

Haraway, D. (1991) *Simians, Cyborgs, and Women: The Reinvention of Nature*, New York: Routledge.

Haraway, D. (1992) *Primate Visions: Gender, Race, and Nature in the World of Modern Science*, London: Verso.

Haraway, D. (1997a) '"Gender" for a Marxist dictionary: the sexual politics of a word' in L. McDowell and J.P. Sharp (eds) *Space, Gender, Knowledge: Feminist Readings*, London: Edward Arnold.

Haraway, D. (1997b) *Modest_Witness@Second_Millennium.FemaleMan(c)_ Meets_ OncoMouse(tm): Feminism and Technoscience*, New York: Routledge.

Harding, S. (1998) *Is Science Multicultural? Postcolonialisms, Feminisms, and Epistemologies*, Bloomington, IN: Indiana University Press.

Harvey, D. (1989) *The Condition of Postmodernity*, Oxford: Blackwell.

Huyssen, A. (1990) 'Mapping the postmodern' in L. Nicholson (ed.) *Feminism/ Postmodernism*, New York: Routledge.

Jameson, F. (1984) 'Postmodernism, or the cultural logic of late capitalism', *New Left Review* 146: 53–93.

Jay, M. (1973) *The Dialectical Imagination: A History of the Frankfurt School and the Institute of Social Research, 1923–1950*, London: Heinemann.

Leonard, S.T. (1990) *Critical Theory in Political Practice*, Princeton, NJ: Princeton University Press.

Leslie, E. (2000) *Walter Benjamin: Overpowering Conformism*, London: Pluto Press.

McChesney, R., Wood, E. and Foster, J. (eds) (1998) *Capitalism and the Information Age: The Political Economy of the Global Communication Revolution*, New York: Monthly Review Press.

McCole, J. (1993) *Walter Benjamin and the Antinomies of Tradition*, Ithaca, NY: Cornell University Press.

Nicholson, L. (ed.) (1990) *Feminism/Postmodernism*, New York: Routledge.

Nicholson, L. (1994) 'Feminism and the politics of postmodernism' in M. Ferguson and J. Wicke (eds) *Feminism and Postmodernism*, Durham, NC: Duke University Press.

Peterson, V.S. and Sisson Runyan, A. (1999) *Global Gender Issues* 2nd edn, Boulder, CO: Westview Press.

Roberts, J. (1982) *Walter Benjamin*, London: Macmillan.

Spiller, N. (ed.) (2002) *Cyber_Reader: Critical Writings for the Digital Era*, London: Phaidon Press.

Weber, S. (1996) 'Mass Mediauras; or, art, aura, and media in the work of Walter Benjamin' in D.S. Ferris (ed.) *Walter Benjamin: Theoretical Questions*, Stanford, CA: Stanford University Press.

Weigel, S. (1997) *Enstellte Ahnlichkeit: Walter Benjamins Theoretische Schreibweise*, Frankfurt am Main: Fischer Taschenbuch Verlag.

Wolin, R. (1994) *Walter Benjamin: An Aesthetic of Redemption*, Berkeley, CA: University of California Press.

Chapter 3

Murray Edelman

Andrew Chadwick

> For most men most of the time politics is a series of pictures in the mind, placed there by television news, newspapers, magazines, and discussions. The pictures create a moving panorama taking place in a world the mass public never quite touches, yet one its members come to fear or cheer, often with passion and sometimes with action . . . Politics is for most of us a passing parade of abstract symbols, yet a parade which our experience teaches us to be a benevolent or malevolent force that can be close to omnipotent. Because politics does visibly confer wealth, take life, imprison and free people, and represent a history with strong emotional and ideological associations, its processes become easy objects upon which to displace private emotions, especially strong anxieties and hopes.
>
> (Edelman 1964: 5)

Political scientists have long sought to explain how political elites maintain themselves in power. This inevitably raises questions to do with legitimation. Even in liberal democratic states, which may have substantial variations in power structures and societal contexts, there is the problem of democratic control; of the relations between rulers and ruled, the relatively powerful and the relatively powerless. Since the late 1950s or so, while the discipline of communication studies blossomed, there have been surprisingly few political scientists who have sought to understand and explain political legitimation with reference to language, symbolism and the manipulation of information; Murray Edelman (1919–2001) is one of them. My aim in this chapter is threefold. First, I map out Edelman's eclectic theoretical influences and analytical framework with the aim of demonstrating how it fits the typically interdisciplinary mould of writing on the information society that is true of the other authors in this book. Second, I tease out its central themes and principal objects of analysis:

language and symbolism. Finally, I translate some of these themes and assess their relevance for understanding political legitimation in the age of the Internet through an empirical analysis of executive branch websites in Britain and the USA. It is my argument that the Internet allows for a new 'electronic face' of government which has previously been unavailable. This is controlled by government itself and is subject to the central demands of early twenty-first-century politics, namely presentational professionalism in the form of attention to imagery, symbolism, language use and genre – all processes to which Edelman's work draws our attention.

About Murray Edelman

Murray Edelman was born in Nanticoke, Pennsylvania, in 1919. He took an undergraduate degree from Bucknell University, Lewisburg, Pennsylvania in 1941, and an MA from the University of Chicago in 1942. Six years later he was awarded his PhD from the University of Illinois, where he stayed until 1966, before moving to a post in the Department of Political Science at the University of Wisconsin-Madison. He remained there until his retirement in 1990, when he was awarded the title of George Herbert Mead Professor Emeritus. As we shall see, this was in recognition of one of his main intellectual influences.

It is fair to say that Edelman's work is more widely known in his native USA than in Europe, although it is probably the case that his first major book, *The Symbolic Uses of Politics* (1964), while it does not feature at the top of the charts in the *New Handbook of Political Science* (1996), is one of the most frequently cited works of its kind in the post-war period (Goodin and Klingemann 1996). Curiously, for a writer who enjoyed such a long career, Edelman's approach remained remarkably consistent. Although touched by the critical influence of Marxism and poststructuralism in the 1980s, his principal works stemmed from essentially the same intellectual paradigm – a mixture of elite theory, social psychology and symbolic interactionism, all of which informed *The Symbolic Uses of Politics* back in 1964. His rather eclectic theoretical approach meant that he was never destined to fit with the dominant behavioural and rational choice tendencies of the US political science establishment. Indeed, many political scientists have long been puzzled by his approach, likening it to social anthropology or psychology rather than conventional political analysis. Murray Edelman died in February 2001, at the age of 81.

Edelman's theoretical framework: a sketch

Extracting the components of a writer's theoretical framework is fraught with difficulties. Not least of these is the problem of contradicting a writer's own views of his/her intellectual debts. To take a relevant example, the suggestive fluidity of Edelman's writing led to condemnation of his 'relativism' by reviewers of his work in the late 1980s (Edelman 1989; Kraus and Giles 1989). To complicate matters, in common with many other political scientists who have nonetheless made major contributions to the discipline, Edelman generally eschewed overtly self-conscious public reflection on how his writings fitted into his intellectual 'biography' in a developmental sense. Thus, the approach I have adopted here derives from the essence of academic writing – that it is the published work of a writer which should stand as a record of their approach. My method has been to examine Edelman's work for its stated and unstated intellectual reference points. In my analysis there are seven distinct yet intersecting and overlapping sources of inspiration for his work that have implications for our understanding of politics in the 'information age'. They are: elite theory; philosophical pragmatism and symbolic inter-actionism; the social psychology of communication; social anthropology; neo-Marxist theories of ideology; poststructuralism; and aesthetics. While all of these were present from his early writings, the influence of neo-Marxism and poststructuralism largely date from one of Edelman's most important books, *Constructing the Political Spectacle* (1988).

The early stages of Edelman's career were, in many respects, the hey-day of political science in the USA. While most writers shared in the dramatic growth of the discipline during the 1950s and 1960s, there were, nevertheless, crucial differences of emphasis. Before western Marxism exerted its influence on US campuses, there was a critical alternative to the mainstream pluralism of writers such as Robert Dahl (1956) and it came in the form of 'elite theory'. Edelman's writing during the 1960s and 1970s can be situated within the set of problems identified by such writers as C. Wright Mills, William Kornhauser and E.E. Schattschneider (Mills 1956; Kornhauser 1959; Schattschneider 1960). The modern elite theorists sought to explain the supreme paradox of liberal democratic politics: why citizens tolerate inequalities in political influence and how elites manage to convince ordinary citizens that this state of affairs is desirable. Mills's *The Power Elite* is particularly important here, not only for its theoretical sophistication, but also for its bridging of the divide between theory and empirical evidence. At the same time, Kornhauser's interpretation of modern politics as based on a division between a knowing elite and an anomic 'mass' – without the mediating civil societal structures identified by pluralists – features strongly in Edelman's writing. But the novelty of the latter's approach lies in its relatively sophisticated elaboration of how symbolism and imagery contribute to elite domination. Mills and others tended to sketch out connections between various sections of the elite – military, social, political administrative or business. Edelman, however, spent little space discussing the empirical detail of elite networks, preferring instead to mix theoretical reflection with observations about language use and symbolic imagery in 'everyday' political contexts, such as bureaucracies, courtrooms and the mass media.

Edelman's understanding of precisely how political support, or 'acquiescence', as he termed it, is manipulated and maintained by elites, has its basis in pragmatism, whose principal figures were John Dewey (1859–1952) and George Herbert Mead (1863–1931). Mead's theories of mind and the self are widely regarded as the foundation of symbolic interactionism (Mead 1930). As Edelman put it, in *The Symbolic Uses of Politics* (1964), Mead's insight 'was his discovery that through a "conversation of gestures" man creates his own world. How people act, which symbols become significant and what they signify, and what there is for men to act upon are not hard "givens", but are created for individual selves through role-taking' (Edelman 1964: 185). Edelman appropriated these abstractions and applied them directly to the symbolic domains of the political in order to understand how meaning and identity is socially embedded through mutually reinforcing acts of communication. Much of this has been given a pessimistic bent, but provides an explanation of how the 'powerful and the powerless cooperate . . . to solidify each other's positions' and how 'symbolic interactions complement economic and social inequalities' (Edelman 1988: 97).

Analysis of how symbolism may be manipulated by political elites to play upon the hopes and anxieties of a mass audience can, in the twentieth century, be traced back to the influence of the British Fabian socialist, and early Professor of 'Political Psychology', Graham Wallas, whose argument about the non-rational basis of politics proved controversial, but undoubtedly influential (Wallas 1908). Edelman also borrowed and adapted reflections on the ubiquity of symbolisation from Suzanne Langer's *Philosophy in a New Key* (1942), as well as Lev Vygotsky's Marxist theorisation of the role played by language in the construction of the self in society (Vygotsky 1962 [1934]). However, the most immediate sources were Harold Lasswell's groundbreaking work on the social psychology of political communications of the 1930s and 1940s (Lasswell 1977 [1930]) and Freudian psychoanalysis. Not only did politicians aim to use the means of communication to play upon irrationality and emotion among the public, argued Edelman, but also both elites and mass were caught up in the inevitability of imperfect communication. Politicians behave in ways that ensure their political survival, and this does not usually mean communicating policy content, but instead rests upon the deployment of symbolic resources and rhetorical strategies. Yet it is political elites who benefit most from the system, since they are able to manipulate mass irrationality for their own ends. From Freudian studies of communication Edelman adapted the notion that individuals crave some kind of sensual and emotional fulfilment from politics once they have lost faith in 'the rationality of social processes'. Thus, when social scientists seek to understand symbols, they must place them in the context of human weakness. In *Constructing the Political Spectacle* (1988), Edelman wrote that symbols 'play their parts only within the context of the hopes and the fears of specific social situations. They reinforce, condense, and reify perceptions, beliefs, and feelings that grow out of such social relations as dominance and dependency, alliance and hostility, anxiety about threats, or anticipation of future well-being' (Edelman 1988: 89). Thus, the weak accept their subordinate social position and adopt the values of their superiors to ensure a form of social 'safety'.

Central to these processes are political myths. In *Political Language: Words that Succeed and Policies that Fail* (1977), Edelman drew upon anthropology, especially the structuralist approach developed by Claude Lévi-Strauss. This provided an explanation of how social policy was dependent upon mythical constructions of the behaviour of 'recipient' social groups, such as the poor, ethnic minorities and women. A mythical 'cognitive structure', the reproduction of which takes place in the mediated public domain, was a central reason for the failure of much social policy. Programmes designed to help those in poverty as a result of unemployment or sickness were bound to fail owing to the pejorative labels applied to such groups by politicians, administrators and the mass media. However, this cognitive structure was crucial for justifying the 'status, power and roles of the middle class, public officials and helping professionals' (Edelman 1977: 8). Edelman argued that political scientists should seek to understand the structured frameworks of meaning which provide the context of policy formulation. This requires sensitivity to how symbolic resources are deployed in institutional power struggles.

Although I have mentioned that Edelman's inception came at a time when the main division in political science was between pluralism and elite theory, by the late 1970s Marxism in its 'neo' or 'western' guise had exerted an influence. There are obvious affinities between his analyses and the theorists of distorted communication and its contribution to social and political inequality, such as Antonio Gramsci, Jürgen Habermas and Noam Chomsky, though, curiously, there are no direct linkages with the Gramscian turn in British sociology driven by figures like Stuart Hall at the Birmingham Centre for Contemporary Cultural Studies in the 1970s. By the time of *Constructing the Political Spectacle* (1988), Edelman was drawing upon Gramsci's theory of hegemony, with its theorisation of the dual pillars of state legitimacy in capitalist society: coercion and consent (Edelman 1988: 103–119). Nevertheless, there was always a central ambivalence in Edelman's treatment of Marxism. There are, to my knowledge, no quotations from Marx in his writings, and the references that do exist are often rather oblique, suggesting unease with an economic determinism that does not sit comfortably alongside his socio-psychological emphases.

This unease perhaps explains Edelman's attraction to poststructuralism in the 1980s, by which time it is evident that writers such as Jacques Derrida, Jean-François Lyotard and Michel Foucault had become significant for his work. In particular, the instability of meaning, the ambiguity of texts and the attention to the creative use of language in power struggles all lent themselves to Edelman's pre-existing disposition towards an anti-realist epistemology (derived from pragmatism) which denied the possibility of a 'world of events distinct from the interpretations of observers' (Edelman 1988: 95). Poststructuralism also provided Edelman with a critique of rational choice theory and other positivist influences in mainstream political science, especially in voting studies and public opinion polling. By this stage Edelman was drawing attention to the ways in which positivist social science in the form of opinion polling was converging with news-as-entertainment. This constituted an intensification of the elite-driven politics he had delineated in his earlier works. He did not go so far as to argue that reality could not be divorced from language, but instead argued that the relationship between language and 'the real' was dialectical, and always dependent upon the context of use (Chadwick 2000). Giving meaning to events through 'labelling' inevitably involved power mechanisms which, in part, shaped language use, but those mechanisms were, in turn, shaped by language itself. In Foucault's approach, individual agency is restricted by the structure which language imposes. This parallels Edelman's view of 'role-taking' derived from symbolic interactionism (Edelman 1988: 112), but, as with neo-Marxism, there is little evidence that the broader 'linguistic turn' in cultural studies, particularly in its British variant, exerted a direct influence on Edelman's work. In many respects,

Edelman's last book, *From Art to Politics*, was his biggest departure from the political science/public policy framework (Edelman 1995). It borrowed much from aesthetic theory, especially the work of Walter Benjamin and Nelson Goodman, in its exploration of how artistic categories influence public life. It is concerned with how genres derived from the world of art influence political perceptions, sometimes in emancipatory fashion, but more often in ways that (again) ensure the dominance of the political elite. Yet the most striking contribution here is the discussion of political 'settings' – the architecture and symbolic lexicon of public and private government buildings, which evoke authority and deference. This built upon an article written for the *Journal of Architectural Education* in 1978, but by 1995 it had been integrated into Edelman's theory of the links between the aesthetic and the political to produce what I consider later on to be a highly suggestive approach to the representation of politics in the symbolic domains of hypermedia. Before that, however, I want to explore Edelman's two enduring objects of analysis in more detail.

Language and the symbolic domains of politics

Given his long career and eclectic theoretical underpinnings, the objects of analysis in Edelman's work remained remarkably constant. His work always focused on two distinct but interrelated areas, both of which have important implications for the analysis of politics in the Internet age: language and symbolic representation.

Since most of us experience politics in a mediated form most of the time, this inevitably introduces the role played by language in shaping perception. While this has always been the case, the intensification of mediated communication caused by the emergence of the Internet raises interesting questions about how political actors may be able to make new and different uses of political language. As a form of communication (at least until the widespread adoption of broad-band connectivity) the Internet places huge burdens on the communicative power of the written word. The main emphasis in political communications literature since the 1960s has been on exploring the impact of television on citizen perceptions and cognition. The grand narrative, although obviously contested, has been about the displacement of older, supposedly superior forms of text – especially newspapers – by the oversimplified, image-centric medium of television. The curious nature of the Internet – with its mixture of words and images – points towards a new, more complex system of political mediation in future, not only because it will coexist with existing forms of media for some time, but also because it might ultimately mean the convergence of existing media forms. In short, it requires us to bring language back in – to see its use and manipulation in a potential digital 'meta-media' as fundamental to future political communication.

Edelman offers us a useful critical tool-kit. He begins from the premise that 'it is language about political events rather than the events themselves that everyone experiences' (Edelman 1977: 142). Thus, 'political language is political reality; there is no other so far as the meaning of events to actors and spectators is concerned' (Edelman 1988: 104). It follows that political scientists must pay attention to the ways that language constructs political reality in different contexts.

In his earlier work, Edelman focused on different linguistic styles and their deployment by political, administrative and judicial elites. Making a distinction between 'hortatory', 'legal' and 'bargaining' styles, he argued that each plays a part in maintaining political support. I have discussed elsewhere how hortatory language is useful for understanding the role of rhetoric in maintaining popular support. The key point here is that linguistic content and form combine to reassure the public that they are being 'consulted' on policy. Hortatory language style – the most common form of political rhetoric, with its use of hyperbole, personalised narratives and appeals to a large audience – serve to construct meaning. The 'meaning' of hortatory political language can be found simultaneously in its substantive content, which varies from case to case, but more importantly in its conveyance of the idea that the public are being appealed to and consulted on matters of public concern; that they are not being marginalised but are central to the political process (Chadwick 2000: 297–298). As I shall illustrate, this language style, albeit in a modified form, is rapidly becoming characteristic of the 'electronic face' of government in the UK and the USA.

These ideas were further developed in Edelman's later work. In *Political Language* (1977), he drew attention to the public's need for reassurance that their political leaders are 'coping' with difficulty: 'This psychological process explains why every regime both encourages public anxiety and placates it through rhetoric and reassuring gestures' (Edelman 1977: 147). The language used by state officials gives the appearance of help, but may be disciplinary and restrictive. The long-term result was policy failure masked by the appearance of policy success (Edelman 1977: 146).

These themes were given a firm empirical grounding in an analysis of political news, which appeared in 1988. By this stage, Edelman had been influenced by poststructuralism, and although he does not use the term, it is the 'intertextual' nature of news which features strongly. Using this anti-realist approach, it is argued that news is based upon multiple layers of interpretation:

> For any audience, then, an account is an interpretation of an inter-pretation. An adequate analysis would see it as a moment in a complex chain of interpretations, each phase of the process anticipating later interpretations and helping to shape them. Ambiguity and subjectivity are neither deviations nor pathologies in news dissemination; they constitute the political world. To posit a universe of objective events is a form of mysticism that legitimises the status quo because the interpretation that is defined as objective is likely to reflect the dominant values of the time.
>
> (Edelman 1988: 95)

The political spectacle, as mediated by news, is dialectical. It both incor-porates and excludes. It is at once a vehicle for maintaining popular support and a reminder of the powerlessness of the public, who only occupy the role of spectators. Similar in function to religious ceremonies, which convey the inaccessible might of the deity, news evokes a sense of political importance through strategic use of language. As Edelman puts it:

> A set of frequently used terms also helps induce an acquiescent posture toward the acts of public officials. Words like 'public', 'official', 'due process of law', 'the public interest', and 'the national interest' have no specific referent, but induce a considerable measure of acceptance of actions that might otherwise be viewed with scep-ticism or hostility.
>
> (Edelman 1988: 98)

The extent to which power relationships in politics emerge out of the symbolic properties of particular sites of interaction, or 'settings', as Edelman termed them, has often been ignored in mainstream political science. Other disciplines, such as anthropology, cultural studies and geography, to name a few, have not shared this neglect. If political scientists are to make sense of the impact of the new ICTs, they need to analyse how the new symbolic domains actually work in shaping meaning and perception. Edelman's theory offers us some useful pointers here, but it requires that we borrow some of his observations about the 'physical' worlds of public space, architecture and the ceremonial and apply them to the domain of computer-mediated communication. One might refer to the latter as the 'virtual', as a means of counter-posing it against the 'real'. However, it is my argument that if we pay attention to the symbolic characteristics of politics in hypermedia, then we are essentially conducting an examination of the same dynamics in both domains. In other words, the symbolic representation which occurs in the form of political settings in the 'physical' world of buildings and ceremonies also occurs in the explicitly political zones of the Internet. Similar processes are at work in the 'face' of government in the 'virtual world' as are at work in the 'physical world'. As citizens increasingly come to interact and participate via electronic means (alongside established physical means), the legitimising role of government's 'electronic face' will assume great significance. But how can we begin to make sense of this phenomenon?

Drawing upon the work of Mead, Langer, Freud and Lasswell, Edelman argued that political life was based in large part upon condensation symbols. In other words, symbols have no intrinsic meaning but come to have significance as a result of what people believe; they condense a range of hopes, fears and emotions. The symbolic settings of politics are, for Edelman, never neutral; they organise and structure the types of action that it is possible for participants to pursue. These symbolic domains condition expectations about how one ought to behave in certain contexts. Thus,

> The courtroom, the police station, the legislative chamber, the party convention hall, the presidential and even the mayoral office, the battleship or chamber in which the formal offer of surrender in war is accepted all have their distinctive and dramaturgical features, planned by the arrangers and actors in the event and expected by their audiences.
>
> (Edelman 1964: 95)

Political settings are usually staged, contrived and even artificial. They often have a 'heroic quality', and are designed to signify 'massiveness, ornateness, and formality' to a large audience (Edelman 1964: 96). This allows them to function as extraordinary, dramatic spectacles which are constructed as intrinsically important, though their outcomes may lack any significance for substantive policy. Nevertheless, there must be a conjuncture between political actions and their contexts for such symbolism to work effectively. For example, the symbolic resources required for a court to function – robes, the bench, chambers, hushed tones and scholarly language – differ radically from those required in a battlefield, where urgency is conveyed by inattention to such ceremonial matters. Individuals are expected to behave differently in different symbolic contexts. Settings also influence individual psychology. Grand architectural imagery evokes in individuals a sense of belonging, of being part of a long-standing, stable order (Edelman 1964: 109). At the same time, however, the symbolic features of the large bureaucratic structures in the modern state act as a barrier to welfare claimants and those seeking information and help (Edelman 1964: 111). If political elites are successful in using such symbolic resources they are able to legitimise their actions.

The most useful and elaborate expression of this approach came in 'Architecture, Spaces and Social Order', published in 1995 as an expanded and updated version of an essay which originally appeared in the *Journal of Architectural Education* (Edelman 1995, 1978). These processes are most acute if we consider the symbolism that surrounds political executives. The architectural spaces within which executive actions are carried out are typically grandiose enough to symbolise 'clarity, order and predictability', 'the power of the presidency and its reflection of the public will . . . reason, merit, or science' (Edelman 1995: 75). They symbolise continuity in a world of flux. Their (literally) 'monumental' character makes them different from their typical surroundings, indicating to their spectators that the inhabitants are inherently powerful, and indicating to their inhabitants that they are marked out as different from the outsiders. Thus, monumentality provides sustenance for both elites and non-elites.

Yet there is a dialectical element to this phenomenon. As Edelman contends, while public buildings demarcate, they also include. We are taught to believe in 'legislative halls, courtrooms, executive mansions, and even administrative offices as symbols of government by the people and equality before the law' (Edelman 1995: 77). It is also crucial to distinguish between the different 'faces' of the State. While the Supreme Court may evoke justice and fairness, the massive headquarters of the Federal Bureau of Investigation (FBI) symbolise – again dialectically – the need for a large state security apparatus to fend off internal subversion, or, for the more critical, the unaccountability and arbitrariness of the modern US federal government (Edelman 1995: 84). While public buildings present themselves in these ways, the internal processes of modern bureaucracies are themselves symbolic representations of the efficiency and remoteness of the government machine. The use of information technology to process information about individuals creates a perception of humans as data to be circulated and manipulated at will. Nowhere is this more obvious than in the welfare system (Edelman 1995: 86).

Finally, Edelman's thinking about symbolic representation dealt with the ways in which electronic media contribute to a politics based upon images and narratives taken from other, often 'entertainment-led' genres, such as films, novels and art. Artistic genres are seen as short cuts to political understanding. They create and manage audience expectations, reducing the amount of 'work' texts have to perform. While art has an influence on how we view politics, at the same time, political elites use references to artistic genres as a legitimation strategy. For example, in *From Art to Politics*, Edelman discusses the impact of kitsch in evoking nostalgia and sentimentality (Edelman 1995: 29–33). As we shall see, government's web presence is not insulated from this form of communication.

The electronic face of government in the Internet age: two examples

Having outlined the principal features of Edelman's contribution to understanding the role of language and symbolism in politics, I now want to suggest briefly how some of these ideas may be used to illuminate recent developments in 'e-government'. I have chosen to analyse what is arguably the most important component of the electronic face of any government: the website of its executive branch. One of the earliest and most successful examples (judged in terms of user numbers) is the US presidential site (http://www.whitehouse.gov). When Internet use began to take off in the USA during the mid-1990s, the White House site rapidly emerged as a first port of call for those seeking information about government. Its perceived ease of use and quasi-portal characteristics proved attractive. By contrast, it was not until relatively recently that the British prime minister's site (http://www.number-10.gov.uk) began to assume the same functionality and popularity. The site was completely redesigned in 2000, and given an intriguing 'brand identity'. It has now emerged as one of the most popular government sites in the UK, and offers a much wider range of content than its US counterpart. It is curious that we often refer our students to executive websites as an information resource, but there has been relatively little critical analysis of their form and content. Much more energy has been spent to date on party websites and election campaigning. But given the symbolic (and very real) power of the executive, even in the most self-consciously liberal democratic political systems such as the USA, an examination of their electronic face reveals some potentially significant aspects of how political legitimation is reinforced through new ICTs.

At the time of this research, the US presidential site was undergoing a period of reconstruction owing to the arrival of George W. Bush. It is

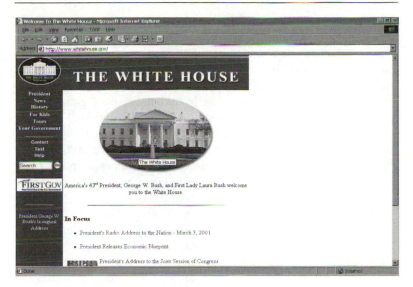

Figure 3.1 The White House web page

likely that the site will soon develop along the same lines as its British
counterpart, though the structural and constitutional differences between
the two states mean that the presidential site is always likely to be less of
a focal point than the British equivalent. In the USA, despite the monumentality and continuity conveyed by the White House site, the separation
of powers also carries some symbolic weight. It is perhaps fitting,
therefore, that the executive branch site is less ambitious. In Britain, where
the new Labour government's intensification of the pre-existing tendencies
towards spin-doctored, sound-bite politics is now such a taken-for-granted
part of national life that it is hardly commented on at all, the Strategic
Communications Unit controls the executive's web presence, and it
shows.

The White House homepage is presented in terms that draw upon the
established appeal of the architecture of the 'real' building (see Figure
3.1). A large, oval-shaped photograph of the front of the White House,
obviously the subject of simple three-dimensional graphics manipulation
to make it stand out, is the most prominent feature of the page, indeed,
of the whole site. Underneath is the phrase: 'America's 43rd President,
George W. Bush, and First Lady Laura Bush welcome you to the White
House'. Traditional colours from the US flag serve as the scheme for
the site, with dark blue predominating over red and white. In this case the
government's identity is established by reference to long-standing motifs:

the monumentality of the building itself, the recognisable colours. These are all artefacts from the pre-web, even pre-photography era. While the overall feel of the site is uncluttered and 'clean', it is certainly not influenced by the genres of hypermedia 'modernism' that inform the design of many corporate sites, and, by way of contrast, the British prime minister's site. The element of tradition provided by the building photograph, and the more obviously official logo, which bears the inscription 'The White House, Washington', is also reinforced by the use of the Times New Roman font (which is coded to override a browser's default font) – a typeface which, though much-used on the web, first appeared in *The Times* newspaper in 1932, and has its origins in the late sixteenth century (Adobe Systems Incorporated 2001; MyFonts.com 2001).

The site is divided into seven main sections: 'President', 'News', 'History', 'For Kids', 'Tours', 'Your Government' and 'Help/Contact'. There is also a prominent link to George W. Bush's Inaugural Address, and a logo link to Firstgov, the US government 'portal' launched in 2000. The latter is again branded in blue, white and red, although a visitor to that site would soon notice a radical difference of style, with Firstgov having a much more consumer-oriented, quasi-corporate design. For the White House site, a sense of contemporaneity, but, crucially for my analysis, executive competence, is provided by a series of news bullets, which appear under an 'In Focus' heading. The audience is invited to click to the full stories, which appear with carefully selected action photographs. Full texts of selected speeches appear, as do audio streams. These stories, at the time of writing, narrate the president acting in some way – either ceremonially through the attendance of the events, or in more concrete terms as the initiator of tax reform policy in one case ('President Releases Agenda For Tax Relief – February 8, 2001'). Some of these are, in Boorstin's classic terminology, 'pseudo-events', but their status is more complex if we consider that the website's strategy, which, as we shall see, differs markedly from the equivalent in Britain, is to legitimise the presidency through reference to long-established motifs and actions 'in the real world' (Boorstin 1992 [1961]). What we are seeing here is a further layer of mediation being added to previously mediated events. Here are Edelman's themes of intertextuality, monumentality and the need for public reassurance that government is actively 'doing something' delivered in the electronic face of the US government. Also much in evidence is the use of hortatory language. Although this comes in the form of reported speeches delivered at external events rather than as direct content designed for the site itself, it is also there in the form of appeals for feedback. Hortatory language forms convey the idea that the audience is being listened to, is being incorporated into the political process, irrespective of their content.

Images of the president do not appear on the site's homepage. An obvious location would be in place of the image of the White House building, but this would detract from the sense of historical continuity built up by the page, and would also undermine the distinction between office and person which is central to political legitimation in liberal democracies. The site signifies that presidents occupy the White House as a symbol of their status and power, but their position is contingent. The audience 'enters' the White House by clicking on links, and 'finds' the president 'within'. Presidential power, history and tradition are therefore condensed in the symbolic representation of the White House itself. The site's domain name reflects this. After all, why not choose 'president.gov'? Upon linking to the president's page, the audience is presented with a colour photograph and an extraordinarily detached mini-biography, narrated in the third person, which tells of Bush's political career before becoming president. Policy and beliefs feature, but are not particularly prominent. The narrator describes how Bush is 'ushering in the responsibility era in America'. The detached written style of the page again reinforces the distinction between occupant and role, and the notion of the White House building as symbolic of continuity and the website as a symbolic extension of that theme. Similar pages exist for the vice-president, Dick Cheney, for the First Lady and for the wife of the vice-president, Lynne Cheney. All are narrated in the third person. Monumentality, continuity, tradition and executive competence are thus conveyed through the form and content of the site.

An examination of the British prime ministerial site, http://www.number-10.gov.uk, reveals similarities and some important differences (see Figure 3.2). The distinction between occupant and role is similarly maintained, with the prime minister's official residence forming the leit-motif. The audience encounters the prime minister sitting within the cabinet section of the site – a nod in the direction of collective leadership. The established symbolism surrounding 'Number 10' is invoked, with a recurring image of the building's famous front door, complete with a smiling police officer. This obviously symbolises the authority and stability of the state, but it is combined with a 'friendly' smile – a strategy of mixing genres which may also be found on many corporate websites. The traditional symbols of the British state, the lion and the unicorn, are transformed into a logo which appears at the top of each page, while the inscription 'Welcome to 10 Downing Street' appears in Times New Roman – the font of familiar historical authority.

Figure 3.2 10 Downing Street web page

Yet these indicators of history, authority, power and status, sit alongside forms and content of a rather different kind. The redesigned Number 10 homepage is the entrance to a hugely expanded collection of different types of content. The main sections are 'Newsroom', 'Magazine', 'Facts', 'Broadcasts', 'Your Say' and a children's section, '10 out of 10'. The style of the site is eclectic, and has obviously been much influenced by contemporary web design, particularly with its use of small, iconic images, animated buttons, different types of font and an overall colour scheme which is not a reference to established, easily recognisable national colours, such as the red, white and blue of the British flag, but a mixture of black, beige, white and green. In other words, the site, though it makes reference to established 'real world' symbols, appears as a distinct entity in its own right; as a product of some reflection on what the web as a distinctive medium can provide rather than the 'brochureware' approach of the White House site. In all, Number 10 is a 'slicker' production, but it is also one that is the product of an obsession with presentation and 'modernisation' that has emerged as a key component of the 'new' Labour government's strategy. Indeed, the site's 'look and feel' borrows heavily from the hypermedia modernism that is characteristic of 'cutting-edge' web design. Contemporary colour schemes, which differ from section to section, iconic 'lifestyle' representations, and clean lines and well-spaced presentation convey an image of vitality and modernity.

Nevertheless, Number 10 is less overt in its portrayal of executive action and competence than the White House. News bullets feature prominently on the Number 10 homepage, but they cover a more diverse range of activities and policy areas, and, at the sample time of this research, none of the stories portrays the prime minister 'acting alone' in the same manner as his US equivalent. Instead, a variety of policy initiatives featuring a range of cabinet ministers are highlighted. The verbal style of the news stories is, at first glance, akin to the pillar of British public service broadcasting – BBC News. But linking deeper into the stories themselves soon reveals that they are little more than government press releases, with their characteristic features: 'The Government has unveiled its vision for a modern, efficient criminal justice system'; '£35 million pilot for pupil learning credits'; 'Courts to get new powers to tackle persistent juvenile offenders'. Thus, executive competence is reinforced, but the collective leadership of the cabinet system is the dominant approach. Competence is demonstrated through coherence and coordination; the Number 10 site is therefore an important element in producing an image of government unity.

Clearly the aim of the news bulletins is also to ensure contemporaneity. Indeed, the portal characteristics of Number 10 make it appear to have been designed as a user's browser homepage. This may be an excessively optimistic assumption on the part of the prime minister's press office. However, its significance should not be underestimated. The wide variety of audience experiences available on the site make it much more likely that users will click around rather than click through, as is likely to be the case with the White House site. Governments have always been in the business of self-publishing, but the web makes it much easier to reach a mass audience with news items that would otherwise have to be channelled through the media. The site offers a customised email update service for users who register, and, at the time of this research, provided a quickly updated information section on the farming crisis caused by the outbreak of 'foot and mouth' disease. Government engaging with 'ordinary' individual web surfers as a direct information provider is a new development, and one that is more likely to be achieved with Number 10 than with the White House as it currently stands. Hortatory language styles predominate on Number 10, just as they do on the White House site.

Both of these executive sites do, of course, feature elements that are electronic translations of the average tourist experience. The White House site has its historical essays on the building and its previous incumbents; Number 10 has similar sections. Visitors to the 'real' buildings would be treated to the same thing. However, what is striking, particularly about the British site, is the extent to which lifestyle content, most of which borrows heavily from glossy magazine genres, is intertwined with political forms and content. Thus, the 'Tour of the Rooms' at Number 10 is reminiscent of an aspirational interiors magazine, and even provides panoramic technology which allows the audience to 'look around' the rooms. Yet the aesthetic description of the rooms is mixed with selected historical observations about their former occupants, safely sanitised down to the level of building materials, but still symbolically powerful. Consider this extraordinary paragraph taken from the description of the Cabinet Room:

> The 23 chairs are the same ones used by Gladstone and Disraeli in the reign of Queen Victoria. Of the set, only the Prime Minister's chair has arms. Some of the silver on the table was presented to the house by President Ronald Reagan. A solid gold sword, presented by the Emir of Kuwait, rests near the window. Those windows are now made of glass three inches thick – a precaution taken following the 1991 mortar attack which shattered the glass.

The '1991 mortar attack' was conducted by the Irish Republican Army (IRA), and came at the depths of the Northern Ireland crisis of the early 1990s. The White House makes similar cultural raids upon established lifestyle and entertainment genres, but it is less pronounced, and mainly revolves around the children's section, 'For Kids'. Nevertheless, the portrayal of the assumption of stable family relationships permeates the site, and constitutes awareness that many adults may be shown the site by their more Net-literate children. In both cases, what can be observed here are classic instances of entertainment genres bleeding into political genres, in much the same way as Edelman describes in *From Art to Politics*. The overall effect is to deflect attention from the reality of executive power and channel the presentation of government into 'safer' areas.

   If entertainment genres help to produce a sanitised version of government, then this is only intensified by the presence of the kind of kitsch that can be found all over the Internet in 2001. The cult of the family homepage, with its ubiquitous sections on household pets, is very much alive and well in the electronic face of government. Visitors to Number 10 are invited to find out about its famous animal occupants, past and present. As for the White House, the Clintons' celebrity cat, 'Socks', has been replaced by the Bushs' rather less charismatic creatures. Taking pleasure in self-publication has long been one of the most popular uses of the Internet, and government is not immune from its characteristic practices. The former British 'e-envoy', Alex Allan, when briefly in charge of Labour's e-government drive, was happy to link from his government pages to his and his partner's 'personal' homepage (http://www.whitegum.com), with its 'Grateful Dead Song and Lyric Finder', Holly the dog, and picture of him windsurfing on the River Thames (Allan 2001).

The executive websites of two of the most advanced countries (in terms of Internet use) reveal the potential for governments to become self-publicists in ways that have previously been unavailable. Disintermediation in the economy is being mirrored in politics. The symbolic architecture of a government's Internet presence is likely to be just as important in the future as it has been in the past, but the emerging techniques point to a more complex relationship between rulers and ruled, one that will be based upon immediacy of contact, a more direct appeal to lifestyle concerns and entertainment values. It has often been argued that the Internet will empower citizens by providing access to information, and there is no doubt that the sheer volume of government information now available online is immense. But aside from structural issues like the 'digital divide', the underlying dynamics of elite driven politics are not going to change overnight, if at all. Hortatory language, for instance, characteristic of political leaders seeking to establish a link with their audience, is both intensified and curiously modified in the electronic face of government, because the citizen actively seeks information. The relatively (though never completely) passive consumption of political language is replaced by a process in which the citizen becomes an active pursuant. Yet there must be information to pursue, and this is controlled and filtered by government itself. Citizens are brought 'closer' to government through their online discoveries, but their interactions with its electronic face are very much on government's own terms.

These sites, especially the British prime minister's, are also typical of the infotainment genres which are fast becoming the stock-in-trade of the more commercial frontiers of the web. That governments are now able to exploit audience recognition of such genres is indicative of the dialectical nature of legitimation identified by Edelman. We are at once in awe of government, and are keen to see the symbolic representation of its power and competence. But we want government to be 'ours' and 'like us'. We want our lifestyle interests reflected and our craving for information and entertainment satisfied by government websites in much the same ways as we would any other site.

Conclusion

The Internet has spawned a new electronic face of government, but one which exhibits many of the features identified as central to political legitimation by Murray Edelman. My analysis of the symbolic forms and content of the UK prime minister's site illustrates a likely future direction for e-government. The rather amateurish, patchy and utilitarian websites of the late 1990s are quickly being replaced by a more professional approach, which ties in with broader government communications strategies. This is undoubtedly a product of the general increase in Internet usage among electorates, but it also represents an increasing awareness of the properties of the web as a medium and how this may contribute to the symbolic dimension of government activity.

The explosive growth of the Internet since the mid-1990s has undoubtedly had an impact on the conduct of politics. But we are only just beginning to appreciate the inevitable balance between continuity and change. Governments will always need support, and the maintenance of support is not always dependent upon rational calculation and the electoral mechanism. Legitimation is a process. It is ongoing, and elite strategies mutate over time. It has been my argument here that much of what Edelman has been writing about since the early 1960s is highly relevant for the Internet age. This is not, however, to state that there is 'nothing new under the sun'. Equally, I am not arguing that the content of legitimation strategies are constant, and that it is only their forms that have changed. This would be to miss one of Edelman's central points – that form and content cannot be separated in some arbitrary fashion. The Internet is a new medium, with properties which undoubtedly draw upon other media, but which, when melted together, make up something rather different. Edelman's work, if suitably 'borrowed', informs us that the Internet offers political elites many opportunities to intensify and diversify the ways in which they sustain themselves in positions of power. The challenge for social scientists is to interpret and explain how these trends may undermine attempts to use new media to reduce political inequality.

One of the more curious features of Edelman's writing was his relative inattention to the structural determinants of communication and how the coming together of technological and economic forces condition the forms of political mediation. This is an obvious criticism of his work, and one that could easily be made by those working within a political economy paradigm. Undoubtedly Edelman's neglect of economic structures means that he had relatively little to say about the historical development of technological forms, nor did he write explicitly of the domination of corporate structures in contemporary media. It is, however, possible to turn this kind of criticism around, and argue that the major weakness of 'materialist' analyses of technology and the media is their relative inattention to the fine-grained psychological, symbolic and linguistic forces which must be understood if we are to assess the importance of any communication process. Indeed, it is possible to go further still, and defend Edelman on the grounds that his work may be situated within broadly the same problematic as the British writer, Raymond Williams, whose 'cultural materialism' speaks of a world in which economic, cultural and ideational entities all have material consequences (Williams 1977). Borrowing from Murray Edelman to understand politics in the Internet age certainly requires some work of translation. But doing so, in the manner I have demonstrated here, makes it possible to imagine that even in cyberspace there will always be a symbolic architecture of power.

Acknowledgements

I thank Christopher May, Rodney Barker and two anonymous referees for their helpful comments and suggestions. Any errors or shortcomings are, of course, my own.

References

Adobe Systems Incorporated (2001) Adobe Type Library: Times New Roman. Available: http://www.adobe.com/type/browser/P/P_145.html (27 February 2001).

Allan, A. (2001) Homepage. Online. Available: http://www.whitegum.com (5 March 2001).

Boorstin, D. (1992) [1961] *The Image: A Guide to Pseudo-Events in America*, New York: Vintage.

British Prime Minister (2001a) 10 Downing Street website. Online. Available: http://www.number-10.gov.uk (5 March 2001).

British Prime Minister (2001b) 10 Downing Street: the Cabinet Room page. Online. Available: http://www.number-10.gov.uk/default.asp?PageId=68 (5 March 2001).

Chadwick, A. (2000) 'Studying political ideas: a public political discourse approach', *Political Studies* 48: 283–301.

Dahl, R. (1956) *A Preface to Democratic Theory*, Chicago: University of Chicago Press.

Dolbeare, K. and Edelman, M. (1985) [1971] *American Politics: Policies, Power and Change*, 5th edn, Lexington, MA: DC Heath.

Edelman, M. (1960) 'Symbols and political quiescence', *American Political Science Review* 54: 695–704.

Edelman, M. (1964) *The Symbolic Uses of Politics*, London: University of Illinois Press.

Edelman, M. (1967) 'Myths, metaphors and political conformity', *Psychiatry* 30: 217–228.

Edelman, M. (1971) *Politics as Symbolic Action: Mass Arousal and Quiescence*, Chicago: University of Wisconsin-Madison Institute for Research on Poverty.

Edelman, M. (1974) 'The political language of the helping professions', *Politics and Society* 4: 295–310. Also reprinted in Edelman (1977).

Edelman, M. (1977) *Political Language: Words that Succeed and Policies that Fail*, New York: Academic Press.

Edelman, M. (1978) 'Space and the social order', *Journal of Architectural Education* 32 (November): 1–9. Also revised and reprinted in Edelman (1995).

Edelman, M. (1980) 'Law and psychiatry as political symbolism', *International Journal of Law and Psychiatry* 3: 235–244.

Edelman, M. (1988) *Constructing the Political Spectacle*, Chicago: University of Chicago Press.

Edelman, M. (1989) 'Alternative interpretations of the political puzzle: a reply to Kraus and Giles', *Political Psychology* 10: 527–531.

Edelman, M. (1995) *From Art to Politics: How Artistic Creations Shape Political Conceptions*, London: University of Chicago Press.

Goodin, R.E. and Klingemann, H. (eds) (1996) *A New Handbook of Political Science*, New York: Oxford University Press.

Kornhauser, W. (1959) *The Politics of Mass Society*, New York: Free Press.

Kraus, S. and Giles, D. (1989) 'Review of Murray Edelman constructing the political spectacle', *Political Psychology* 10: 517–525.

Langer, S. (1960) [1942] *Philosophy in a New Key: A Study in the Symbolism of Reason, Rite and Art*, Cambridge, MA: Harvard University Press.

Lasswell, H. (1977) [1930] *Psychopathology and Politics*, Chicago: University of Chicago Press.

Mead, G.H. (1930) *Mind, Self and Society*, ed. C.W. Morris, Chicago: University of Chicago Press.

Mills, C.W. (1956) *The Power Elite*, New York: Oxford University Press.

MyFonts.com (2001) Times (New) Roman. Online. Available: http://www.myfonts.com/Article18.html (27 February 2001).

Schattschneider, E.E. (1960) *The Semi-Sovereign People*, New York: Holt, Rinehart & Winston.

United States President (2001) The White House website. Online. Available: http://www.whitehouse.gov (27 February 2001).

Vygotsky, L. (1962) [1934] *Thought and Language* (trans. E. Hanfmann and G. Vakar), Cambridge, MA: MIT Press.

Wallas, G. (1908) *Human Nature in Politics*, London: Constable.

Williams, R. (1977) *Marxism and Literature*, Oxford: Oxford University Press.

Chapter 4

Jacques Ellul

Karim H. Karim

The views of Jacques Ellul (1912–1944) on propaganda and myth have significant potential for understanding the ideological promotion of information society in the present. While he is often cited on issues of technology and globalisation,[1] Ellul's unique observations on the role of myth in technological society reside in general obscurity. The references to him by political economists are usually in the context of the autonomy of technique, sociologists generally talk about his work on alienation, political scientists about the role of the state in technological society, communication scholars about propaganda, and theologians about morality.[2] Studies of myth in contemporary life have also generally failed to draw from Ellul's insights on this topic. He contextualised, within the twentieth century, the age-old nature of myth as a basis for providing meaning. The French sociologist presented a loose network of myths that can usefully serve as frameworks for analysing the manners in which technology is conceptualised and promoted presently. His passionate exploration of propaganda in technological society demonstrated the ways in which technique is presented as a means to attaining utopia, akin to the relationship between religious ritual and paradise.

Whereas this sociologist is best known in scholarly circles for *The Technological Society* (*La Technique ou l'enjeu du siècle* 1954)[3] and *Propaganda* (*Propagandes* 1962), a large proportion of his other work deals with theology and ethics. He was primarily concerned with the issue of maintaining moral values in contemporary civilisation. *The Technological Society* and *Propaganda*, both works of social philosophy, reflect his concern with the spiritual condition of human beings in the twentieth century.

A brief scan of Ellul's life helps put into context his particular intellectual approach to technological society. He was born in Bordeaux on 6 January 1912. In his long and varied career he was a member of the French

Communist Party during the mid-1930s, a fighter for the French Resistance during the Second World War, the Deputy Mayor of Bordeaux between 1944 and 1947, and a professor at Bordeaux University's law school and its institute of political studies later in life. By the time he died in 1994 he had published forty-three books and numerous articles.

This thinker's concern with the spiritual bases of life is noteworthy considering that he found religion relatively late in life, at the age of 22. His faith evolved out of the Death of God movement and the response of the neo-orthodox theologians Bultmann, Barth, Niebuhr and Tillich. Ellul's active commitment to Christianity led to membership in the National Council of the Reformed Church in France. He was also a consultant to the Ecumenical World Council of Churches from 1947 to 1953. However, his ecumenism did not extend far beyond Christianity.[4] Much of his corpus is overtly Christian, but a few of his works including *The Technological Society* and *Propaganda* adopt an approach that can be described as spiritual humanism. In the translator's introduction to *The Technological Society* John Wilkinson notes that Ellul sought to bear witness, in the Christian sense, 'to testify to the truth of both worlds and thereby to affirm his freedom through the revolutionary nature of his religion' (Ellul 1964: xx).

Contemporary relevance

Ellul's work is becoming increasingly relevant in light of the growing interest in exploring the manner in which information society is symbolically constructed. This chapter deals specifically with the French sociologist's discussions of the use of myth by the proponents of technological society and will seek to extend his mode of analysis. Quite distinct from the recent cultural, sociological, economic, political, political economic and philosophic examinations of technology, there is a growing body of literature that examines it from perspectives of myth, spirituality, religion and ethics. Apart from those critics of technological society who have used the notion of 'myth as falsehood' (e.g. Smith 1982), a significant number have explored the concept of myth as a mode of conceptualising the world.[5] They have generally attempted to demonstrate the manipulation of myth to promote the ideological purposes of technological interests.

There is a range of approaches among scholars who have discussed spirituality and religion to describe or critique the relationship of human beings with technology. Rushkoff (1994) and Kurzweil (1999) foretell the merging of the soul and the silicon chip. Davis (1998), Stefik (1996) and Dery (1996) write about the 'technomystical impulses' of New Age 'technopagans' who reassert the ascendency of the spiritual in the face of post-Enlightenment materialism. However, Wertheim (1999) challenges the spiritualising of technology and argues that it cannot support religious world-views. Noble (1997) suggests that the millennialist notions of technological development blind us to our real and urgent needs, and he urges that we disabuse ourselves of the 'religion of technology'. In another vein, Hopper (1991) suggests that the substitution of a humanly constructed future for an other-worldly one has created disillusionment which continual technical innovation seeks to deflect. In this he draws from the arguments of Christian theologians such as Tillich (1988), who criticised the moral and ethical ambiguity of technological civilisation. Sardar (1996) echoes this thought in his critique of western materialism manifested in cyberspace. Mowlana (1996) finds that the information society and Islamic community paradigms have philosophic and strategic conflicts in the conceptualisation of human relationships. And Pannikar's (1984) Buddhism-inspired analysis of western technology suggests that the machine enhances power but diminishes freedom. It is freedom that Ellul sought to reassert in the face of technique.

The Technological Society and *Propaganda* were written within the broad context of mass society, which was the subject of numerous treatises in the early and mid-twentieth century. However, Ellul's work was quite distinct from others such as the Chicago or the Frankfurt School in his particular blending of Marxist and theological thought, which sought to develop a critique of the fundamental sociological engines of contemporary society. He went beyond a mere examination of technology to present an analysis of the basic mode of technique, which he defined as 'the totality of methods rationally arrived at and having absolute efficiency (for a given stage of development) in every field of human activity' (Ellul 1964: xxv). Technique is the overarching feature of contemporary western civilisation, according to Ellul; it describes not only the modalities of technology but also the nature of present-day social, political and economic organisation. The rationalism, functionalism and secularity engendered by the Enlightenment shape technique. It privileges decontextualised, quantitative measures and forms of analysis over qualitative ones, and has led to the erosion of moral values and to the mechanisation of traditional rhythms of life. Political and commercial propaganda (advertising) has taken the place of the debate of ideas.

Ellul held that technique has promoted the atomisation of society and the reciprocal concentration of power in the nation-state. Whereas the status of the individual has been elevated vis-à-vis the group, the human being has paradoxically lost autonomy to technique. Humans have become alienated by living in societal conditions that are bereft of meaning. While Ellul borrowed from Marx, he criticised him for embracing technique as a weapon against the bourgeosie, thus promoting subservience to the greater evil. But rather than seeking to destroy technique, the former's quest was to enable human beings to transcend technological determinism. He hoped that his readers would begin to regain a sense of responsibility and freedom from technique by merely becoming conscious of the control it has over their lives (Ellul 1964: xxxiii).

In this lies one of the major problems with Ellul's work. Struggle against technique-dominated society takes place at the personal level in Ellul's scenario, despite the admission that 'an uncommon spiritual force or psychological education is necessary to resist its pressures' (1964: 377). The absence of structural strategies with which to confront technique may perhaps be an extension of his theological existentialism, says Mowlana, but in placing 'the onus of choice and salvation on the individual' he leaves us 'intelligently unhappy' (Mowlana 1997: 10). And what appears to make the individual's task even more daunting is that technique has sought to colonise even his spiritual endeavours (Mowlana 1997: 415–427). This appears to lead to a dead end – no resolution seems possible; we are only made aware of the problem.

Ellul does not appear to provide a revolutionary way of overcoming the hegemony of technique. The value of his unique contribution, however, is to be found in his extensive exposition of the use of deep myth for ideological purposes. His examination of the ways in which myth was used in the mid-twentieth century has much resonance in the study of contemporary rhetoric about the 'information revolution', 'global information infrastructure' and 'knowledge-based society'. The promotion of new media on a transnational scale by government, industry and media can be usefully studied within the conceptual framework put forward by Ellul. Not only did he prove prophetic in his description of the increasing intensification of technique in various aspects of life, but also his examination of propaganda in technological society offers significant insight into the ways the integrally capitalist vision of 'global information society' is being publicised around the world.

The leaders of industrialised countries speak in glowing (almost religious) terms about the potential of the Internet to create prosperity for all. Al Gore was among the earliest to present a vision of universal access to health care, university education, and employment through a 'National Information Infrastructure' (Government of USA 1993). At the July 2000 Group of Eight summit in Tokyo, one of whose main agenda items was to deal with the enormous debt problems of the poorest countries on the planet, the 'world leaders' magnanimously offered – in the fashion of Marie Antoinette – the dream of development through global electronic networking.

> Indeed, we are in the midst of a worldwide effort, organized by many different companies and governments in many different ways, to make computer communication a transcendent spectacle, the latest iteration in [David] Nye's . . . 'technological sublime'. Everything from advertising to trade shows, from demonstration projects to conferences, speaks of a campaign to market the magic, to surround computer communication with power, speed, and the promise of freedom.
>
> (Mosco 1998: 61)

Industry giants such as Microsoft, Cisco Systems and Nortel promote the view that one merely has to 'build and they will come'; in this they continue to ignore the rapidly growing disparities between those who are able to use information technology for their benefit and those who are not (United Nations Development Programme 1999). The mass media have become panegyrists to the 'new economy', whose chief executives are made out to be superstars among celebrities. *Wired* magazine predicts a future in which 'digital citizens' or 'netizens' will ensure greater democracy and the World Wide Web will become the index for all human knowledge.

The new media are viewed as being the answer to humanity's problems; they will create the perfect society. Viewing technical innovations in this manner is not new, as Marvin (1988) has demonstrated. Utopic visions have long accompanied technological evolution. 'Utopia' was the imaginary island envisioned by Thomas More in the 1500s. Its inhabitants had created an earthly paradise by adhering to a pious life and a communal spirit as well as to technical pursuit. In the following centuries, the Enlightenment gave rise to the notion of utopia that could be attained through continual technological and social development. Bury noted in his landmark study about the idea of human progress that at its basis it

> is a theory which involves a synthesis of the past and a prophecy of the future. It is based on an interpretation of history which regards men as slowly advancing . . . in a definite and desirable direction, and infers that this progress will continue indefinitely . . . It implies that . . . a condition of general happiness will ultimately be enjoyed, which will justify the whole process of civilization.
>
> (Bury 1920: 5)

The idea of progress views continual technological developments leading to increasing comforts, the elimination of disease and the eradication of poverty, i.e. heaven on earth. Even though this was mainly a secular notion, Bury suggested that it drew inspiration from the linear conception of history in dominant Christian discourse. After the expulsion of Adam from the Garden of Eden his descendants had been offered redemption at a particular moment in time by Christ, who would return sometime in the future to lead the faithful back to paradise after ruling on earth for the period of a millennium.

The utopic vision of progress is characterised by a millennialism that conceives of a progressive movement towards ultimate fulfilment. Whereas religion offers individuals the dream of paradise, they are promised material comforts in an earthly utopia in the secular domain. The means of progress for the righteous is good deeds and correctly performed ritual, while for the adherents of the technological world-view it is hard work and the proper use of technology. Contemporary propaganda implies that technological improvements within information society will ultimately lead to the arrival of the perfect state in which all desires of consumers will be fulfilled. Wertheim notes that many characterisations of the Internet are drawn from biblical descriptions of heaven (1999: 256–261).

Most cultures seem to make sense of their temporal existence in terms of the ultimate end of their collective existence. The narrative themes of end-times and of a messiah-like figure are pre-biblical and are to be found in many religions (Thompson 1996: 3–16). Utopic dreams appear to be integral to universal human myth. Even though the particular idea of progress is only a few hundred years old, it fits into ancient ways of thinking. In this, the age-old millennial myth is integral to contemporary discourse about information society (Karim 2001).

Myth

In Greek, the word for 'myth' denotes 'narrative'. The limited scope of this chapter does not allow for elaboration upon the numerous views on the nature of myth. In accordance with Ellul's use of the term, this essay views myths as providing frameworks for various sets of meaning – they help make sense of what we experience. A key issue that I would like to address is whether myth is historically determined or whether it has essential, unchanging and universal characteristics. The mythological lores of various peoples are indeed shaped by their respective histories, cultures and environments. Whereas some mythic narratives make sense across cultures, there are others that do not. Certain stories may cease to be of value after centuries of forming the basis of a world-view. But then there are also those particularly resilient mythic themes that seem to recur over time in different forms around the world (Campbell 1968).

In Ellul's understanding of myth it is impossible to create an entirely new myth (Ellul 1965: 199). It is only elaborations of existing myths that emerge from time to time. Ellul indicates that there are certain myths, e.g. that of paradise (1964: 191), that form the basis of others; although he does not provide us with a consistent system of the evolution of myths. Nevertheless, this perspective opens up the possibility of tracing the origins of sets of myths to certain central myths of humankind. We can conceive of some core myths as providing meaning for the basic aspects of human consciousness.[6] These first-order myths include notions of the self, the other, time, space, knowledge, creation, destruction, causes and effects.[7] These fundamental cognitive frameworks seem to be ahistorical and integral to human existence.

They give rise to a second order of myths (see Table 4.1): that of the self to kinship, community, nation, race; the other to gender, nature, divinity, enemy; time to history, beginnings, endings; space to distance, geography; knowledge to learning, science, wisdom, gnosis; creation to life, birth, rebirth; destruction to death; causes to actions, work; effects to reward, punishment, and so on. These second-order myths do not form exclusive frameworks of meaning; they may overlap or interact with each other sporadically or continually. Many of them are shared universally; however, large numbers of people no longer adhere actively to the myths of divinity, gnosis or afterlife, and those of gender, nation, race and distance are the subject of much debate. Second-order myths therefore appear prone to historical change. From them emanate other levels of myth, which I shall discuss later.

Jacques Ellul's attempt to outline some of the basic myths used in technological society's propaganda seems more promising than Barthes' approach in helping us understand the relationships between myth-based messages in information society rhetoric. Viewing myth from the perspective of semiotics, the latter is primarily concerned with its ideological tendencies rather than its nature as a fundamental receptacle for meaning. Indeed, Barthes tends to conflate myth and ideology. To him myth's 'function is to distort' (1973: 121). This limited view of myths permits him to study them only as fleeting phenomena, all of which are subject to history. Barthes provides some fascinating analyses of the ideological dynamics of myth in popular culture, advertising and news, demonstrating its ability to depoliticise, dehistoricise and naturalise ideological messages.

Table 4.1 First- and second-order myths

First order	Second order
self	kinship
	community
	nation
	race
other	gender
	nature
	divinity
	enemy
time	history
	beginnings
	endings
space	distance
	geography
knowledge	learning
	science
	wisdom
	gnosis
creation	life
	birth
	rebirth
destruction	death
causes	action
	work
effects	reward
	punishment

But, even though he is sometimes cited in studies addressing the role of myth in the promotion of information society, his approach is limited in providing the basis for a thorough-going examination of the topic.

Ellul, on the other hand, draws to our attention the fundamental cognitive power of myths whose origins lie at the basis of human consciousness. He also provides a useful distinction between myth and ideology, and attempts to explain their interactive relationship. However, he does not present a systematic method for his analysis of myth, and in places even seems to contradict himself. Nevertheless, it is possible to apply the broadly based conceptual scheme that he used in his study of the mid-twentieth-century civilisation to the study of the contemporary information society.

Ellul's understanding of myth is influenced by his conceptions of spirituality. For him, human beings pay attention to myth because it responds to their sense of the sacred. The very secularisation of society has

heightened this sense because people have an essential need for a continuing connection with spirituality. This need is fulfilled ironically by
technological society's magnificent achievements, which fill the individual
with a sense of wonder.

> He is seized by sacred delirium when he sees the shining track of
> a supersonic jet or visualizes the vast granaries stocked for him. He
> projects this delirium into the myth through which he can control,
> explain, direct, and justify his actions . . . and his new slavery.
>
> (Ellul 1964: 192)

Technique does not eliminate religious tendencies but subordinates them
to its own purposes. In displacing spirituality, technique itself becomes an
object of faith. It comes to embody the sense of mystery that was once the
province of religion (Ellul 1964: 141–142). This is manifested presently
in the mass amazement expressed towards the capabilities of the Internet;
it seems magical, even miraculous, in enabling activities that were
supposedly impossible. The superlatives attached to the capabilities
of electronic networking are generally considered to be beyond question,
quite like religious attitudes to divinity. Uncritical consumption of media
reports with selective data, that show the remarkable progress engendered
by technique, becomes part of the mystical devotion to 'the power of
fact' (Ellul 1964: 303), which is characteristic of post-Enlightenment
empiricism.

In *Propaganda*, Ellul dwelt extensively on the manner that propaganda,
as technique, is used to mobilise masses. The function of propaganda, he
maintained, is to provoke action. It does this by arousing mythical belief
because myth incites strong emotions in people. Pre-propaganda continually feeds audiences with mythical narratives, creating psychological
readiness. Education, advertising, movies, magazines, the discourse of
public institutions and the modes of technology operate collectively as
sociological propaganda to create conformity. Even intellectuals are not
immune (Ellul 1965: 76). Finally, active propaganda, which directly and
vigorously draws on myth, moves people to action in a specific direction.
The effects of communication have been debated for many decades; Ellul
resided in the camp that held that propaganda does work.

The evidence of propaganda's effects is to be found not only in totalitarian states but also in western democracies, which use the myths of
peace, freedom and justice to ensure obeisance to the ruling elite (Ellul
1965: 243). Democracy, which is supposed to foster the expression of
opinions, itself has been turned into a myth. Mass participation, which

was a feature solely of religion, has been adopted by democracy: 'anti-mystique becomes . . . mystique' (1965: 244). In this, democracy becomes religious: 'The content of this religion is of little importance; what matters is to satisfy the religious feelings of the masses; these feelings are used to integrate the masses into the national collective' (1965: 251). Within this context, Ellul suggests that the 'American myth . . . [presents] exactly the same religious traits as the Nazi or Communist myth' (1964: 422).

Religious beliefs, which provide answers for the unknown, enabled human beings to face fear. They also provided meaning for the vagaries of life and death, and fostered a sense of community. In secular society, propaganda functions in a similarly totalising manner. It provides meaning for everything for the inhabitant of technological society and addresses his 'most violent need to be reintegrated into a community' (1965: 148). The creation of community through electronic communication is a primary theme in information society discourse. As propaganda takes the place of religion, it leaves the individual incapable of engaging in critical thought. 'In this respect, strict orthodoxies always have been the same' (1965: 167).

The French sociologist sought to distinguish myth from ideology. He viewed ideology, in its basic form, as 'any set of ideas accepted by individuals or peoples, without attention to their origin or value' (Ellul 1965: 116). It also has the characteristics of cherishing particular ideas, relating to the present, and being dependent on belief rather than proven ideas. Myth differs from ideology in that it

> is imbedded much more deeply in the soul, sinks its roots farther down, is more permanent, and provides man with a fundamental image of his condition and the world at large. Second, the myth is much less 'doctrinaire' . . . The myth is more intellectually diffuse; it is part emotionalism, part affective response, part a sacred feeling, and more important [*sic*]. Third, the myth has stronger powers of activation, whereas ideology is more passive (one can believe in an ideology and yet remain on the sidelines). The myth does not leave man passive; it drives him to action.
>
> (Ellul 1965: 116)

In order to motivate people, ideology necessarily has to link itself to myth. In so doing it grafts contemporary political and economic elements to the psychological/spiritual force of core human beliefs. Ideology cannot create a new myth – it is forced to ally itself to an existing one, which it then strives to elaborate in the propagandist's preferred direction (1965: 119–200). Therefore bourgeois ideology interacted with the already-existing Christian

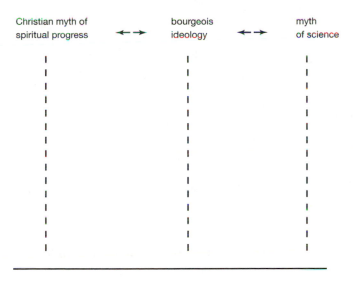

| Christian myth of spiritual progress | ← → | bourgeois ideology | ← → | myth of science |

Contemporary myth of technological and social progress

Figure 4.1 Ideology and the myth of progress

myth of (spiritual) progress and the Enlightenment's myth of science to give rise to the industrial myth of (technological and social) progress (see Figure 4.1). This hybrid myth was vital in harnessing mass society to the project of large-scale industrialisation in the nineteenth and twentieth centuries. The more recent myths of information society and globalisation have been derived from similar interactions, but more on that later.

Ellul also distinguishes myth from another psycho-sociological motivator, namely, the 'collective presuppositions' of technological society. He identified these as the following:

• Everything is matter
• History develops in endless progress
• Man is naturally good
• Man's aim in life is happiness

(Ellul 1965: 39)

These collective presuppositions are 'a collection of feelings, beliefs, and images by which one unconsciously judges events and things without questioning them, or even noticing them' (1965: 39). Myth goes much deeper than these tacit areas of social consensus:

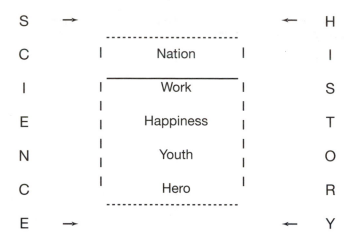

Figure 4.2 Operation of primary and secondary technological myths

It is a vigorous impulse, strongly colored, irrational, and charged with all of man's power to believe. It contains a religious element. In our society the two great fundamental myths on which all other myths rest are Science and History. And based on them are the collective myths that are man's principal orientations: the myth of Work, the myth of Happiness (which is not the same thing as the presupposition of happiness), the myth of the Nation, the myth of Youth, the myth of the Hero.

(Ellul 1965: 40)

Here we have what I have termed above as second order myths, which have for the most part predated technological society but which are used in a particular manner by its propagandists. The myths of work, happiness, nation, youth and hero do not emerge from those of science and history, but operate within their dialectical locus (see Figure 4.2).

A framework for analysis

Ellul does not dwell at length on the interaction between the myths of science and history. However, an extensive commentary on this process is to be found in George Szanto's *Theater and Propaganda* (1978). The latter's use of Ellul's work on myth indicates how it can be effectively used as a framework for analysing contemporary society. Szanto proposes that the primary myths of science and history reflect contrary ideals and

serve society best when they function in dialectical interaction . . .
implicit in the myth of Science is the concept that the absolute in
human improvement can be achieved, the utopia can be built; the
myth of History suggests that so long as man is based in a material
universe his condition will continue to change and the concept of
improvement must be modified and developed for each new generation.

(Szanto 1978: 40)

He asserts that propagandists of the 'liberal/conservative hegemony'
primarily use the myth of science in order to keep audiences passive
and satisfied. Dominant discourses of the technological state are based
upon the interpretations of the myth of science in order to maintain
consensus about the continuing viability of the status quo. The myth of
history, on the other hand, is kept invisible from the masses as it views
all situations as continually changing and changeable. And the secondary
myths of the nation, work, happiness, youth and the hero – which can
be viewed through either the myth of history or the myth of science –
are mainly presented by liberal/conservative integration propagandists
through the latter primary myth. Basing the secondary myths on the
myth of science appears natural and logical to the citizen of the technological state, whose epistemological outlook has been conditioned by
the predominantly rationalist, empiricist and positivist discourses of the
Enlightenment.

The idea of the nation-state was developed by late-eighteenth- and
nineteenth-century western political theorists also working in accordance
with the myth of science. It was an attempt to establish workable geographical bounds to which the individual mind could attach its loyalty and
within which it could visualise the achievement of its goals. Even under
globalisation the nation-state remains entrenched; indeed, its legislative
and bureaucratic structures are essential for legitimising the creation of
'global information society'.

The myth of globality, which is central to contemporary dominant
discourses, does not feature in Ellul's work, much of which was conceived
prior to the current period of intensive globalisation. This second-order
myth emerged from the dialectic between the core myths of the self and
the other, and is akin to that of humanity.[8] The myth of globality grew
with the round-the-world sea voyages, the colonisation of other continents
by Europeans, the growing acceptance by anthropologists of the concept
of the one human race and the increased ease of global communications;
pictures from outer space of the 'one world' reinforced visually the notion

Table 4.2 Myths related to globalisation

First order	Second order	Third order	Fourth order
self	(humanity)	(universal church) (universal *ummah*)	
other	globality	globalisation	global information society; global free trade; elimination of national borders; global triumph of democracy

of its inhabitants' interdependence. The myth of globality has given rise to a third-order myth, that of globalisation, which is historically specific to our age (see Table 4.2). Its earlier manifestations are to be found in the Christian concept of the universal church and that of the universal Muslim *ummah*. In turn, from this myth has emerged a fourth-order myth of global information society and others such as global free trade, the elimination of national borders, the global triumph of democracy, and so on.[9] These have all emerged from ideological imperatives specific to technological society.

Ellul stated that the five secondary technological myths of the nation, work, youth, happiness and the hero continually interact with each other in the discourses of integration propagandists. He saw the myth of the nation as providing the socio-political parameters within which the secondary myths can operate. The national economy becomes the primary frame for the myth of work, national security for that of happiness, and service to the nation for those of youth and the hero. Increasingly, the myth of globalisation is coming to share this role with that of the nation.

The myth of work, operating within the myth of science, presents the cost and the functional efficiency of commercial goods and services as the most important economic characteristics. Interpreted through the myth of history, it would emphasise, from a Marxist point of view, the labour of the worker (Szanto 1978: 41–45) and, from a religious point of view, the spiritual elements of human activity (Shariati 1979: 97–110). The myth of history would also consider relevant the uses of the product, making the worker as well as the consumer participants in the historical process and in helping determine the future of their respective local and global environments. However, dominant discourses operate continually to subvert alternative modes of conceptualisation that would threaten hegemonic structures. They use the myth of work together with the myth of the nation (and that of globalisation) to portray the untroubled functioning of the systems of production and consumption as the correct and normal state of being in technological society. Interruptions such as labour strikes or consumer boycotts are usually portrayed in the mass media as aberrations. The dominant myth of work would have all members of society go about performing their assigned tasks within the national and global systems of production and consumption as if functioning like a giant clockwork machine.

Ellul identifies the nineteenth-century bourgeois world-view as the source of contemporary conceptions of the morality of work:

> Work purifies, ennobles; it is a virtue and a remedy. Work is the only thing that makes life worthwhile; it replaces God and the life of the spirit. More precisely, it identifies God with work: success becomes a blessing. God expresses his [*sic*] satisfaction by distributing money to those who have worked well.

(Ellul 1964: 220)

The second-order myth of work, drawing on the core human myth of causes, is interrelated with that of happiness, which is dependent on the core myth of effects (see Figure 4.3). These two are linked by the myth of

	history		
	cause	→	effect
religious world-view	good deeds	→	paradise
			happiness
technological world-view	work	→	earthly utopia
information society rhetoric	labour of 'knowledge workers'	→	comfort through consumption of the information economy's products

Figure 4.3 Progress as the path to paradise/utopia

progress (Ellul 1965: 117), which is derived from that of history. The myth of happiness is also closely linked to another second-order myth, that of utopia/paradise, which promises the ultimate reward for hard work/good deeds. Intellectual labour in the 'knowledge-based society' is presented as leading to a happy and comfortable life through the consumption of the very materials produced by 'knowledge workers'.

Szanto (1978) sees the passive myth of happiness and the active myth of youth as remnants of late-eighteenth- and early-nineteenth-century romanticism that glorified individualism. These two myths engage in struggle within the person: 'Although the myth of Happiness implies a desire for security and comfort (Science), or resolution and synthesis (History), it can nonetheless live side by side with the myth of Youth: exploration and discovery, potential, the future, risk for great reward' (Szanto 1978: 46). The inner creative conflict in the individual caused by the two secondary myths is used by the integration propagandist to conjure the illusion of escape from technological alienation and the constraints of political control.[10] Apart from their dyadic interaction, the myths of happiness and youth also play a complementary function within the pentad of the secondary myths, particularly with that of the nation. Myths of globalisation tend to take a back seat in propaganda of this nature. The mass media, adhering to dominant political discourses, glorify the zest of youth and its pursuit of happiness usually within the context of national laws and norms. Youth can find ultimate happiness through service to the state, which is presented as the ideal territorial locus for the actualisation of utopia.

Military service to the state is glorified in the mass media mainly through depictions of selfless dedication of young men engaging in combat against overwhelming odds. Television advertisements for recruitment into the armed forces exploit the myth of youth, offering 'exploration and discovery, potential, the future, risk for great reward' (Szanto 1978: 46). Military facilities are often made accessible for commercial film companies to produce what are little more than feature-length advertisements for various branches of the armed forces. They show young soldiers, sailors and airforce personnel using highly sophisticated armaments with accomplished skill and bravado; in accordance with the primarily scientific interpretation of the myth of youth, the answer to the world's problems is to be found through high-technology weaponry, not in the continual re-evaluation of socio-political circumstances through the myth of history. Media highlighting of the much-vaunted accuracy of computer-controlled 'smart bombs' fired by young people flying the most advanced military jets stands out as a key example of this tendency. Popular lore also encourages the notion that children whose hand–eye coordination becomes proficient in playing computer games will be the ace fighter pilots of tomorrow.

The myth of the hero, which is also often narrated in the context of service to the state, is a vital tool in keeping the masses pacified. In the dominant discourse of western literature, the hero, generally a young, white male, usually comes from outside to regenerate a decaying society (Szanto 1978: 49–52). After having saved the community, the hero either moves on or settles down into it. Both these patterns work well for the integration propagandist's purposes: they serve to further the illusion that an individual in the form of a white knight can alleviate a society's problems. According to Szanto:

> The propagandist will usually admit that the society is not perfect. But he will claim it is perfectible, a priori within the myth of Science, and the shining knight, through individual action, can save or improve or ameliorate or cleanse the society, and bring it closer to perfection. The propagandist nurtures the myth of the Hero because he knows it is impotent in fact but powerful in image – he knows *that the individual alone can never alter the economic base of the capitalist state*.
>
> (Szanto 1978: 50, original emphasis)

The hero cannot be allowed to be outside the control of the hegemony: the hero's image is shaped in accordance with the needs of the integration propagandist.

However, within the myth of history the hero is organically united with the community, operating in the framework of practical realities. The hero works within the context of the group, leaving it 'only to attain the historical perspective needed by the group, returning to it constantly' (Szanto 1978: 51). Hegemonic political discourses, which legitimise particular ways of viewing public problems, discourage this form of hero from gaining prominence in the public mind since the hero will be seen as operating in accordance with the requirements of the masses. They will also prevent the hero from appearing in opposition to the state: such a position is reserved for the 'anti-hero' or villain (often the role played by the terrorist or the computer hacker, or the hybrid 'cyberterrorist').

In the New World, the myth of the hero has been linked even more strongly than in Europe to that of the nation: this person is visualised in the context of the frontier pioneer pushing the bounds of the pale, denoting the eternal possibility of extending the borders under the sway of civilisation.

When the contemporary nation can no longer expand its territory its corporate heroes continue broadening the scope of their activities extra-territorially. This is where the myth of globalisation is pressed into service: the White House's vision of the 'National Information Infrastructure' (NII) was enlarged into the 'Global Information Infrastructure' (GII). In earlier periods, other myths based on the fundamental myth of the self had been used to justify colonising alien lands (i.e. extending the self) in the names of civilisation and Christianity. These had interacted with the myths of progress and paradise, as does the contemporary myth of globalisation.

Indeed, paradise becomes the ultimate goal and justification of techno-logical civilisation's propagandists. 'Complete satisfaction is guaranteed' for those who totally integrate into information society. Advertisers tell them that they will have the jobs they want, the products and services they wish for, the personal life they seek, the body shape they desire and the lifestyle of their choice – all for the price of an Internet subscription. The armchair explorer of cyberspace, as contemporary frontier pioneer, finds Eldorado in the World Wide Web. Popular culture often portrays the contemporary hero as physically merging with technology; television characters such as 'Bionic Man', 'Bionic Woman', 'Robocop' and *Star Trek*'s 'Data' as well as the potential for genetic manipulation of human beings is fuelling visions of the new superman and superwoman. In the larger sense, technology itself is the real hero or the messiah/saviour who will reign over the earthly paradise.

Conclusion

The implications of information society's propaganda are enormous. Its ultimate aim seems to be complete absorption of everyone into a perfectly working system of production and consumption that benefits only a few. It conflates data and information with knowledge and wisdom, promising a paradisical state of happiness for all who plug into the Internet. Several governments have decided that since this medium seems to be beyond control it can be self-regulated by industry. Many leading politicians appear to have staked their personal futures on the success of the infor-mation economy. It is touted as the answer to the problems of society. Whereas the moral debate on the Internet appears to be limited to content such as pornography and hate literature, the broader ethical challenges of unequal access, the differential opportunities for use, and the growing gap between those who are able to take advantage of the new media and those who are not go largely unaddressed. The myths of progress and of paradise are vital in explaining the vaunted benefits of the information society for all its inhabitants. This propaganda works because these myths draw on the fundamental modes of human cognition.

A re-evaluation of Jacques Ellul's work in the context of the role of myth in information society helps to challenge its rhetoric at the fundamental level. His work reintroduces an appreciation of the age-old roots of contemporary discourse on technology. Studies that seek to understand myth-based appeals in information society rhetoric will not be able to comprehend the strength of such discourse without taking into account their deep psychological/spiritual elements. Ellul's assertion that it is possible to trace most social myths of technological society to the myth of paradise urges us to enquire into the ways in which new myths are generated from ancient founts. His attempt to describe the manners in which propagandists exploit existing myths to develop attachment to emergent ideologies provides insight into current developments. The view that various myths operate in interactive networks allows us to construct analytical frameworks that trace the emergence and use of particular myths towards specific ends. This also enables the development of counter-strategies that challenge fundamental bases of dominant ideologies. A rediscovery of the French sociologist's work on myth, propaganda and technological society therefore permits a fresh perspective into the discursive construction of the information society. It also encourages the broadening of scholarly activity beyond the strictly positivist confines of dominant social science.

Notes

1 Cf. Rajaee (2000), Noble (1997), Feenberg and Hannay (1995) and Babe (1990).
2 For critical reviews of Ellul's corpus see Christians and Van Hook (1981) Fasching (1981), and Lovekin (1991).
3 He rewrote and expanded that book and published it under the title *The Technological Bluff* (*Le Bluff technologique*, 1988).
4 Ellul's views seem to have been inimical towards Muslims, judging from his *The Betrayal of the West* (1975) and *FLN Propaganda in France during the Algerian War* (1982) as well as his preface in Bat Ye'or's *The Dhimmi* (1980).
5 Babe (1990), Ferguson (1992), Hamelink (1986) and Mosco (1998, 1999).
6 Whereas core myths may be called 'concepts', I call them 'myths' in order to underline the conscious mind's inability to produce purely objective ideas. Even fundamental myths are the result of human attempts to make sense of our environment and experiences; they enable us to interpret the world and our place in it. However, this is not to imply that every human conception can be placed under the rubric of myth; Ellul himself distinguishes myth from ideology and 'collective presuppositions', as is discussed later.

7 These first-order myths are not meant to correspond to those outlined by
   Barthes (1973).
8 Both Christianity and Islam adhered to the concept of a universal humankind.
9 Marjorie Ferguson, while conflating globality with globalisation, lists 'seven
   myths about globalization': 'Big is Better', 'More is Better', 'Time and Space
   Have Disappeared', 'Global Cultural Homogeneity', 'Saving Planet Earth',
   'Democracy for Export via American TV' and 'The New World Order'
   (Ferguson 1992: 74).
10 Ellul asserts that the technological society's propagandists present the
   individual's happiness as the ultimate justification for its activities (1964:
   390).

References

Babe, R.E. (1990) *Telecommunications in Canada: Technology, Industry and
   Government*, Toronto: University of Toronto.
Barthes, R. (1973) [1957] *Mythologies* (trans. A. Lavers), London: Paladin.
Bury, J.B. (1920) *The Idea of Progress*, London: Macmillan.
Campbell, J. (1968) *The Hero with a Thousand Faces*, Princeton, NJ: Princeton
   University Press.
Christians, C.G. and Van Hook, J.M. (1981) *Jacques Ellul: Interpretive Essays*,
   Chicago: University of Illinois Press.
Davis, E. (1998) *Techgnosis: Myth, Magic + Mysticism in the Age of Information*,
   New York: Harmony.
Dery, M. (1996) *Escape Velocity: Cyberculture at the End of the Century*, New
   York: Grove.
Ellul, J. (1964) [1954] *The Technological Society* (trans. John Wilkinson), New
   York: Vintage.
Ellul, J. (1965) [1962] *Propaganda: The Formation of Men's Attitudes* (trans.
   K. Kallen and J. Lerner), New York: Alfred A. Knopf.
Ellul, J. (1978) [1975] *The Betrayal of the West* (trans. M.J. O'Connell), New
   York: Seabury.
Ellul, J. (1982) *FLN Propaganda in France during the Algerian War*, Ottawa: By
   Books.
Ellul, J. (1985) [1980] 'Preface' in B. Ye'or, *The Dhimmi: Jews and Christians
   under Islam*, London: Associated University Presses.
Ellul, J. (1990) [1988] *The Technological Bluff* (trans. G.W. Bromily), Grand
   Rapids, MI: Eerdmans.
Fasching, D.J. (1981) *The Thought of Jacques Ellul: A Systematic Exposition*,
   Toronto: Edwin Mellen.
Feenberg, A. and Hannay, A. (eds) (1995) *Technology and the Politics of
   Knowledge*, Bloomington, IN: Indiana University Press.
Ferguson, M. (1992) 'The mythology about globalization', *European Journal
   of Communication* 7: 69–93.

Government of USA (1993) *The National Information Infrastructure: Agenda for Action*, Washington, DC: Government of USA.

Hamelink, C.J. (1986) 'Is there life after the information revolution?' in M. Traber (ed.) *The Myth of the Information Revolution*, Beverly Hills, CA: Sage.

Hopper, D.H. (1991) *Technology, Theology and the Idea of Progress*, Louisville, KY: Westminster/John Knox Press.

Karim, K. (2001) *Myth and Symbol in Technological Society*, London: I.B. Tauris/Institute of Ismaili Studies.

Kurzweil, R. (1999) *The Age of Spiritual Machines*, New York: Viking.

Lovekin, D. (1991) *Technique, Discourse, and Consciousness: An Introduction to the Philosophy of Jacques Ellul*, London: Associated University Press.

Marvin, C. (1988) *When Old Technologies were New*, Oxford: Oxford University Press.

Mosco, V. (1998) 'Myth-ing links: power and community on the information highway', *Information Society* 14: 57–62.

Mosco, V. (1999) 'Cyber-monopoly: a web of techno-myths', *Science as Culture* 8(1): 5–22.

Mowlana, H. (1996) *Global Communication in Transition: The End of Diversity?*, Thousand Oaks, CA: Sage.

Mowlana, H. (1997) *Global Information and World Communication*, Thousand Oaks, CA: Sage.

Noble, D.F. (1997) *The Religion of Technology: The Divinity of Man and the Spirit of Invention*, New York: Alfred A. Knopf.

Panikkar, R. (1984) 'The destiny of technological civilization: an ancient Buddhist legend Romavisya', *Alternatives* 10: 237–253.

Rajaee, F. (2000) *Globalization on Trial: The Human Condition and the Information Civilization*, Ottawa: International Development Research Centre.

Rushkoff, D. (1994) *Cyberia: Life in the Trenches of Hyperspace*, San Francisco, CA: HarperCollins.

Sardar, Z. (1996) 'alt.civilization.faq: cyberspace as the darker side of the west' in Z. Sardar and J.R. Ravetz (eds) *Cyberfutures: Culture and Politics on the Information Highway*, New York: New York University Press.

Shariati, A. (1979) *On the Sociology of Islam* (trans. H. Algar), Berkeley, CA: Mizan.

Smith, A. (1982) 'Information technology and the myth of abundance', *Daedalus* 111(4): 1–16.

Stefik, M. (1996) *Internet Dreams: Archetypes, Myths, and Metaphors*, Cambridge, MA: MIT Press.

Szanto, G. (1978) *Theater and Propaganda*, Austin, TX: University of Texas.

Thompson, D. (1996) *The End of Time: Faith and Fear in the Shadow of Time*, London: Sinclair-Stevenson.

Tillich, P. (1988) *The Spiritual Situation in our Technical Society* (ed. J.M. Thomas), Macon, GA: Mercer University Press.

United Nations Development Programme (1999) *Human Development Report 1999: Globalization with a Human Face*, New York: United Nations Development Programme.

Wertheim, M. (1999) *The Pearly Gates of Cyberspace: A History of Space from Dante to the Internet*, New York: Norton.

Chapter 5

Harold Innis

Edward Comor

> Innis . . . offers a poetic polysemic discourse that is impenetrable to reason. He is sanctified as the first of Canada's post moderns, a bricoleur whose output requires not rational assessment but aesthetic appreciation or Kabbalistic decoding by a contemporary priesthood of connoisseurs and cultists.
>
> (Collins 1989)

> Innis's concern lay in the thought processes through which people of different civilisations define their vision of reality . . . [H]is focus is less on the individual than on the character of the society that produces individuals and either releases or suppresses their creative potential.
>
> (Cox 1995)

> Why do we attend to the things to which we attend?
>
> (Broeke quoted in Innis 1982)

Fifty years after his death, the body of work produced by Harold Adams Innis (1894–1952) remains widely cited but frequently misunderstood by students of communication studies (Acland and Buxton 2000). Born at the end of the nineteenth century in south-western Ontario, Innis is most certainly Canada's most prodigious social scientist. Predating and directly shaping his work on communication, Innis was an internationally recognised political economist and historian. Through his early interest in markets, related social-historical structures, and the role of transportation networks in relation to them, Innis's interest in communication and culture began. Most famously, Innis introduced his concept of media bias in a 1935 paper on the intellectual habits of social scientists (Innis 1935) many years before his explicit studies on communication. It was only in his later years, particularly after being diagnosed with cancer, that Innis developed and applied bias in what became an increasingly obsessive effort to draw scholarly attention to the dynamics underlying the general inability of twentieth-century western civilisation to redress its cultural imbalances.

Innis never considered his concept of bias to be some sort of analytical fulcrum through which the causal mysteries of history could be revealed tout court. Nor, as Collins believes, should his work be classified as some kind of subjectivist mantra useful only to a 'priesthood of connoisseurs and cultists' (Collins 1989: 218). Instead, as Cox recognises, bias and other Innisian concepts were developed as heuristic tools to help us better understand those forces and relations shaping society's critical and creative capacities (Cox 1995: 20, 28). McLuhan, in his later 'Introduction' to Innis's 1951 publication *The Bias of Communication* (Innis 1982), adds that 'Innis taught us how to use the bias of culture and communication as an instrument of research' (Innis 1982: xi). Indeed, it is my view that Broeke's philosophical question – 'Why do we attend to the things to which we attend?' – prominently quoted by Innis in this book, should be understood as the defining question in Innis' communication studies.[1]

This chapter argues that misreadings of Innis's work in general, and his concept of bias in particular, to some extent affirm the reason for Innis's initial formulation of bias – a concept first used in an attempt to enable social scientists to be explicitly reflexive. In what follows, I shall explain Innis's concept of bias in both the context of his larger body of work and in terms of its contemporary relevance. In the first section, common-place criticisms or misinterpretations of the bias of communication will be addressed and clarified. The second section focuses on Innis's more general methodology involving a form of dialectical materialism that is overtly concerned with the dynamics of power, how people think, and the long-term implications of technologies, organisations and insti-tutions in relation to these. Here, a heuristic model is presented as a means of summarising Innis's work. The third section applies both bias and this model to questions regarding the potentials and implications of Internet-based technologies in the early twenty-first century. The final section – the conclusion – underlines Innis's political concerns with contemporary developments and the overwhelming cultural bias he observed – an orien-tation towards spatial dominance and away from temporal sustainability. It argues that in light of capitalist-based globalisation and technological developments, these concerns are perhaps more pressing today than at any time in history.

The bias of communication

Even in Canada, where Innis was born and worked (at the University of Toronto), his *Bias of Communication* (1982) was not immediately well received. Innis had established himself as that country's pre-eminent social scientist based largely on his 'staples approach' to Canadian economic history and its more general implications for political economy. This work involved Innis in a series of decidedly holistic, materialist and dialectical but explicitly empirical analyses of how frontier economies develop. Through this work, Innis revealed that the ways in which economists had come to understand economic history involved assumptions based on the experiences of relatively developed political economies. Moreover, Innis demonstrated how developments 'at the margins' of the world economy, based largely on the extraction of staple resources, entailed a complex of structural conditions and subsequent political economic dynamics. Through this work, Innis recognised, in extraordinary detail, the inter-connections between various regions and vested interests and the crucial roles played by transportation, communication, and culture in these power-laden relationships (Innis 1995). As such, Innis's apparent turn away from Canadian economic history and his staples approach to his relatively abstract analyses of communication over four thousand years of history – a project conducted in just the final years of his life – seemed to his contemporaries a rather eccentric and less than successful pursuit (Acland and Buxton 2000: 8–10). Despite the apparent suddenness of this 'turn' and subsequent isolation from his many colleagues, in many ways these later studies directly involved many of Innis's earlier concerns and analytical tools. As Innis put it, 'it is part of the task of the social scientist to test the limits of his tools and to indicate their possibilities' (Innis 1982: xvii). Indeed, what had really changed was his use of a much broader historical canvas, his compulsion to emphasise concepts rather than empirical detail, and the explicitly political concerns driving his work both forward and, paradoxically, deeper and deeper into the intellectual wilderness.

What Innis referred to as the 'biases' of core institutions, organisations and technologies – *the nodal points through which what we know and how we know are produced and reproduced* – constituted his core concern throughout these final years. For Innis, a communication medium may facilitate the capacity to control space (or territory) as a necessary prerequisite to increasing control over time. In other circumstances, similar attempts to increase control over space could lead to a decline in the capacity to control time. As explained below, the bias of communication is not a reductionist concept. It is a heuristic tool in which dialectically related contexts are crucial. For Innis, the cumulative effects of how people communicate through a broad range of media, over any given time and at any given place, are not reducible to isolated social or physical characteristics. To apply bias, a comprehensive assessment of history is required in order to identify key media and to generate an elaborated understanding of their influence on history.

Clarifying bias through his critics

To help in the task of explaining bias, it is useful to be clear about what it is not. The acerbic comment by Richard Collins used to open this chapter was published in his review of the re-release of Innis's 1950 book *Empire and Communications* (Innis 1986). In his critique, Collins (1989) criticises this collection of Innis lectures most essentially because they present neither a 'systematic' nor 'well-focused' argument. Because, according to Collins, the 'impact of communications . . . is not sufficiently differentiated from the effect of other factors' on historical development, he believes that Innis fails to show 'that the structure and nature of communications has been any more decisive a force in the life of empires' than factors such as social organisation, legal and familial systems, and military rule (Collins 1989: 217).

The fundamental difficulty of this and similar criticisms is that they assume that Innis shares a popular definition of what 'communication media' are. Collins also assumes that the absence of both precise definitions and the presence of difficult prose constitute little more than 'a set of take it or leave it dogmas . . . camouflaged by a thick frosting of sparkling information – facts lining the nest of an intellectual magpie and concealing the fundamental intellectual disorderliness of Innis's system' (Collins 1989: 218). More to the point, Collins is taking an intellectual stand against the absence of 'a clear structure of argument presenting . . . propositions that are open to testing and selective discard or appropriation' (Collins 1989: 219). In contrast to what are deemed to be his unscientific communication studies, Collins commends Innis's earlier staples studies for upholding this standard. Faced with writings that are 'impenetrable to reason' (thus supposedly breaking 'the rules' of scientific discourse), Collins concludes that 'Innis's later works are weathering badly in comparison to his earlier monuments' (Collins 1989: 218–219).

More common than this complete rejection of Innis's later writings is the tendency to misread and/or misappropriate his work generally and the concept of bias in particular. Some who have done this have labelled Innis a technological determinist. Marvin, for example, writes that Innis 'leaps from technological "fact" to social "effect"' (Marvin 1983: 32). Innis, she continues, 'failed to realize that meaning is not in the technological object, but only in the particular practices to which society puts it' (Marvin 1983: 35). Specifically, Marvin assumes an all too common reading of what Innis meant by the bias of communication:

> Innis uses the term 'bias' to specify media orientation. Time-biased media render the passage of time unimportant in the transmission of messages. However far back in time a message is launched, it remains unimpeded and undistorted. People separated by generations can have the same message in their hands . . . Space-biased media render the expanse of space unimportant in the transmission of messages. From no matter how geographically distant a point a message is launched, it remains unimpeded and undistorted.
>
> (Marvin 1983: 32)

In her reading of Innis, Marvin classifies 'media' in accordance with their space or time 'biases'. Time-binding media include the spoken language, clay, parchment and stone because they are characteristically durable and difficult to transport. Space-biased media, on the other hand, are light and fragile, permitting wide-scale distribution but limiting their duration over time. These include paper, celluloid and electronic signals. According to this reading of Innis, time-biased media foster hierarchy, decentral- isation, provinciality and tradition, whereas space-biased media promote centralisation, bureaucracy, secularism, imperialism and the use of force (Marvin 1983: 32). As Couch summarises: 'Innis . . . sought to demon- strate how the media are social environments *sui generis* that determine broadsweeping everyday forms of social consciousness and social relationships' (Couch 1990: 112).

While these two planes of criticism – one rejecting Innis's communi- cation studies as some sort of postmodernist ruse, and the other critiquing its supposedly obvious reductionism – appear unrelated, they both, in fact, are rooted in a general ignorance of the intellectual heritage of Innis's communication studies in the context of his methodology and related political concerns.

The origins of bias

As already mentioned, Innis first used bias in 1935, five years prior to the publication of his last well-known staples study, *The Cod Fisheries* (Innis

1940). This early application emerged out of Innis's attempt to specify the dynamics that shape the subjective tendencies influencing the work of the social scientist. Rather than a concept developed to prioritise the role of communication in historical development, bias was first developed as a heuristic tool employed in the task of empowering social scientists, encouraging them to develop a reflexive mode of intellectual practice. The paper in which bias is introduced is called 'The role of intelligence in the social process'. It was written in response to an article by Urwick (1935) who argued that the natural science paradigm was not suitable for the social scientist because, unlike the natural world, the social world is inherently unpredictable and ever-changing. This state of affairs, said Urwick, is largely a result of the inherent unpredictability of the thoughts and actions of basically free-willed human beings. Reflecting debates that are very much with us today, Urwick wrote that the social scientist also is inevitably infused with subjectivist tendencies. As such, no human being could hope to be objective while examining and interpreting the inevitably unpredictable subject of social behaviour. 'Life', according to Urwick, 'moves by its own immanent force, into an unknowable future' (Urwick 1935: 76).

Innis both challenged the belief that human behaviour is ultimately unpredictable and Urwick's subsequent rejection of the scientific project. While agreeing that most behaviour is spontaneous and that human beings (including academics) often act on the bases of ingrained behavioural patterns involving degrees of unreflexive thought, Innis countered Urwick by recognizing that *these thoughts and practices are themselves developed and reproduced*. He called these thoughts and practices 'biases' and generally recognized them to be historically determined. Innis thus made an important assertion: while objectivity is impossible, social scientists can develop the analytical tools needed to become aware of their own subjectivities, how they are constructed, and how and why they are unconsciously expressed again and again.

With a touch of tongue-in-cheek, this general point is made by Innis in the following passage:

> The innumerable difficulties of the social scientist are paradoxically his only salvation. Since the social scientist cannot be 'scientific' or 'objective' . . . he can learn of his numerous limitations . . . The difficulty if not impossibility of predicting one's own course of action is decreased in predicting the course of action of others, as anyone knows who has been forced to live in close relations with one other person over a considerable period of time. The exasperating accuracy

with which such prediction is possible has been the cause of more than one murder in northern Canada and the dissolution of numerous partnerships.

(Innis 1935: 283)

Innis goes on to explain that 'the sediment of experience provides the basis for scientific investigation' and that 'the habits or biases of individuals which permit prediction are reinforced in the cumulative bias of institutions and constitute [or should constitute] the chief interest of the social scientist' (Innis 1935: 284).

It is here that Innis establishes the framework for the development of the bias of communication. By examining how day-to-day lives are mediated by organisations and institutions – how the key nodal points of social-economic power affect thoughts and practices – Innis believed that social scientists could and should take preliminary steps in the task of redressing their own biases and their sometimes negative implications for the state of social knowledge.

This concern pushed forward Innis's emerging focus on the role of communication media (broadly defined) in the history of western civilisation. Troubled by the rapid growth of specialisation in social science in the early twentieth century, Innis was concerned that the university itself was becoming the arbiter of instant solutions rather than an essential source of critical questions. After 1945, he observed the dissipation of critical voices in the political culture of the Cold War. In the past, ignorance and a belief in quick solutions could produce military conflict. In the emerging nuclear age, this concoction could well lead to the annihilation of humanity. Such weighty concerns compelled Innis to pursue the aforementioned question, why do we attend to the things to which we attend? Given the mobilisation of weapons of mass destruction and accompanying Cold War paranoia, Innis believed that by applying this question while rereading history – particularly in terms of what he observed to be the dialectic between what he called 'monopolies of knowledge' and 'monopolies of power – social scientists potentially (and, for Innis, perhaps even heroically) could develop the intellectual keys to human survival.

Bias in the context of history and power

By monopoly of power, Innis was addressing the predominance of entities capable of applying extraordinary military resources. By a monopoly of knowledge, he addressed those interests possessing extraordinary control over what information is available and/or those having a predominant influence over more complex patterns or habits of social thought. In other words, this latter 'monopoly' involves explicit and/or implicit control over the social pool of information and how that information is used in developing what is 'known'. As a trained economist (who, near the end of his life, became the first non-American to be appointed President of the American Economics Association), Innis also recognised that both power (i.e. force) and knowledge are directly related to control over wealth.

By the time his first collection of essays that explicitly addressed communication – *Political Economy in the Modern State* (1946) – was published following the Second World War, Innis recognised organisations, institutions and technologies as 'communication media' in that they constitute core structures through which people interact and history itself unfolds. Through this focus, Innis again underlined his concern with the underpinnings of human biases and how they are affected by predominant institutions, organisations and technologies. As such, Innis came to understand the bias of communication directly to affect, and be affected by, those interests engaged in the struggle to control force, knowledge and wealth.

Contemporary interpretations and applications of bias often involve relatively narrow or uninformed readings. More often than not, fragments of Innis's work have been extracted and applied as if they could be read 'straight', without interpretation. In his communication studies many instances can be found in which Innis makes statements concerning the biases of particular technologies. For example, Innis would write that durable media, such as stone, 'emphasise time' and that the use of stone implies a time-biased society. Ancient Egypt constitutes an example of this. Through the use of pyramids and temples, Innis argues that the medium of stone provided the Pharaonic class with the bases for their sustained and long-term dominance.

In raising this technology–power relationship, it is important to point out that a deeper, more 'political' project is being pursued. Through his suggestive rather than empirically detailed mode of presentation, Innis's communication studies actively seek to engage the reader in a kind of dialogue. By focusing on, for instance, the durable character of stone, Innis is addressing only one aspect of the physical capacities of what was then a predominant medium of communication. In his writings, Innis always took pains to use words such as 'emphasise' and 'implies' when referring to bias. To illustrate this further using a medium popularised during his lifetime – the radio – Innis at first considered it to have had tremendous participatory capacities as a result of its potential emphasis on conversation and debate, both in-studio and through its integration with listeners over telephone lines. Based largely on its physical capacities, he recognised that radio could be used to promote the development of democratic exchange and mass critical thought. Innis also understood radio to be potentially positive in terms of its capacity to act as a counter-balance to the largely one-way communication tendencies found in the popular press. Rather than reading centrally produced material, crafted to attract and maintain mass or specialised consumers, the radio presented at least the possibility of generating a thoughtful and socially inclusive dialogue.

Innis, however, understood that the application and impact of radio – as with all communication media – also involved the *context* and, more particularly, the economics of its development and control. As a predominantly commercial medium (at least in the US context), radio, like the press, for the most part became yet another vehicle used by private-sector interests to attract consumers to advertising. Through rigid schedules, well-defined personalities, and the sensual rather than the intellectual engagement of audiences, such mass market commercial priorities only served to deepen the emerging monopolisation of knowledge in twentieth-century North America. While recognising radio as a vehicle through which the predominance of short-term thinking could be redressed (through a very public exchange of ideas and interpretations), the context of its use, particularly in the USA, generated a bias characterised by the predominance of power structures interested in controlling demographic markets and political-economic territories. In the case of radio, for Innis, the context of capitalism most directly facilitated its use to further the already dominant cultural bias of spatial expansion over collective memory and longevity.

All in all, in order to understand both his writings on communication and his concept of bias, Innis must be read in the context of his concern with the very ideal that Collins defends in his critique – the scientific aspirations of the social scientist. The bias of communication and Innis's accompanying study of history were pursued in an effort to advance our understanding of why we attend to the things to which we attend. As a result of the technocratic tendencies and aspirations of most of his academic contemporaries, and the general absence of a critical public able to redress an emerging oligarchy of specialised experts, Innis feared that 'the conditions of freedom of thought are in danger of being destroyed by science, technology and the mechanization of knowledge, and with them, western civilization' (Innis 1982: 190). It is the task of the social scientist, thought Innis, to overcome this cultural bias through the rebalancing of scholarly concerns – away from a search for concrete facts and toward the elaboration of abstract ideas; away from answering questions and more toward the framing of them.

Crises and the dialectics of power

The Bias of Communication (1982) is a collection of essays that apply the concept of bias in a decidedly non-deterministic way. The main goal of its chapters is to use communication media as focal points through which macro-historical developments can be better understood. More specifically, for Innis, the development and implementation of media – involving coinage, the horse, the price system, the university, the radio and innumerable others – signal a response to social and/or economic and/or military uncertainties or crises. In turn the application of a significant new communication medium or set of media itself contributes to the restructuring of the human and natural environments.

For Innis, periods of uncertainty or crisis constitute historical moments in which disturbances in the capabilities held by dominant interests become apparent. Put another way, the apparent decline in the capacity to maintain or expand territorial control and/or maintain control over time signals the need for a reorganisation of institutions and/or organisations and/or technologies. This often involves attempts, as Innis often put it, to establish or extend the monopolisation of knowledge and this involves implicit or explicit efforts to control predominant ways of seeing and thinking. Innis understood that media play important roles in the dissemination of ways of knowing through space and/or time. Efforts to control space and/or time also involve attempts to monopolise force which, according to Innis, involve a range of control activities from brutal oppression to the more subtle implementation of surveillance technologies.

Because Innis believed that the development and implementation of significant new communication media often signal attempts to redress uncertainty or crisis, he thought that the social-economic collapse of historical empires reflects the failure of existing strategies to control space and/or time – strategies that are directly conditioned both by what is known and the ways in which what is known becomes or remains known. By recalling that bias was introduced in his formative staples writings, and its application in reference to ways of conceptualising the world, it becomes apparent that Innis's work consistently is concerned with the capacity of a society to recognise and resolve crises. As Innis warned, '[e]ach civilization has its own methods of suicide' (Innis 1982: 141). A contemporary example of this, our deepening environmental crisis, serves as an illustration.

An Innisian perspective would view the contemporary environmental crisis in terms of the predominant way we see or understand ourselves in relation to the natural world. This involves the presence of an almost ingrained bias, characterised by an obsession with the short-term and a generally acritical approach among most commentators and public officials in relation to the long-term systemic causes of pollution. Most fundamentally, the predominance of particular biases – such as the view that growth and competition are inherently 'progressive' – tend to limit what is culturally feasible or realistic in efforts to respond to this and other crises. As I note later, this malaise is being directly conditioned by contemporary media (e.g. the Internet) and its development in the context of capitalism.

Innis observed that biases tend to be cumulative and self-reinforcing. This is important because what is feasible or realistic today – since it fundamentally reflects the way of thinking that facilitated crisis in the first place – may simply serve to 'hold-the-fort' or 'buy time'. Such 'solutions' also may serve to exacerbate the problem structurally thus making the crisis, over the long term, less rather than more correctable. Owing to the cumulative tendencies and intellectual characteristics of bias, societies often unconsciously construct barriers to the long-term resolution of their systemic problems. Again using the environmental crisis as an example, watered-down versions of the sustainable development paradigm, for instance, become an apparent solution. Through the concept of bias – because it compels the analyst to focus on historically produced and structurally ingrained intellectual habits – this kind of thinking can be recognized and potentially redressed.

Innis's dialectical materialism

As noted previously, bias constitutes just one element in Innis's more elaborate methodology. A related concept is Innis' time–space dialectic. For Innis, throughout history, efforts by a particular group, collectivity or class to assert power, explicitly or implicitly, usually involves problematic efforts to control the temporal and spatial conditions (both mental and physical) of day-to-day life. Through historically structured biased media, powerful concerns often will attempt to normalise their interests as if they were natural, universal, inevitable. Thus, for example, the pyramids of ancient Egypt served the Pharaohs and priestly class by spatially representing their eternal presence and God-like capabilities. Today, as discussed in the next section of this chapter, the Internet, in relation to the context of its development and use, can be viewed as a medium whose moment-to-moment obliteration of *both* spatial and temporal barriers

Figure 5.1 Innis's dialectical triad

serves to normalise (or make 'inevitable') the perspectives of those with vested interests in particular modes of globalisation in relation to those who may not.

From Innis's general body of work, a heuristic model can be constructed involving his implicit conceptualisation of a struggle involving not only time, space and the temporal or spatial biases of predominant media, but also (as discussed earlier) control over knowledge, wealth and force. This model (Figure 5.1) constitutes a means of anticipating and assessing potential developments involving how changing media environments effect power relations.

The struggle to control knowledge, wealth and force can be represented as a dialectical triad that serves to make explicit significant historical tensions and possible contradictions. In relation to bias, this struggle is directly shaped by predominant and historically structured media (institutions, organisations and technologies) at any given place and time and their often unobserved effects on social epistemologies. In the context of this model, hegemonic stability rarely is attained over sustained periods of time and resistance (whether organised or fragmented) constantly plays a role in the outcome of particular tensions and in the restructuring or development of media in the future.

According to this model (Figure 5.1), human beings, their histories and constructions, all take place within the context of the earth's natural environment. As arrows flowing into and out of the middle of the diagram indicate, the ongoing and overarching limitations of nature are themselves subjected to human-generated modifications. The next and, of course, dialectically interrelated level in this model is the predominant mode of production or 'how we produce and reproduce' our collective lives. Through this level, how society at any given place and time organises its material survival – from hunter-gatherer, to slave-based, to capitalist political economies – is recognised to be the next essential context affecting (and affected by) human thought and action. At the centre sits the triad itself. Here it is assumed that a given social or world order involves the predominance of, or struggles involving, the interests of some in relation to others.

Furthermore, relative stability (or hegemony) presupposes the capacity of particular interests to control the interrelated components of power: knowledge, wealth and force. How human beings conceptualise themselves, their world and their interests, in the context of this ongoing struggle, is conditioned by innumerable local, national and global media. As such, history unfolds in the context of our existence in relation to the natural environment, our predominant political economies and the realities (and perceptions of reality) shaping power struggles.

Innis and the Internet

Innis's work provides valuable tools in efforts to assess what has become the focus of great interest some fifty years after his death – the character and implications of the Internet and more general digital technology developments. In Innisian terms, questions related to these include: will such technologies serve to democratise communications, breaking the monopoly of knowledge built up over the twentieth century by mostly large-scale corporate entities? Or, will the context of capitalism and its complementary technological, organisational and institutional mediators suppress such potentials, thereby consolidating the power of capital in deeper and more expansive ways?

As discussed, through his historical research, Innis believed that the development or significant reform of media communication takes place, most typically, as a response by vested interests facing some kind of crisis in their capacity to control knowledge and/or wealth and/or force. In the twentieth century, Innis believed that time and again (with radio being his most contemporary example) the full potentials associated with communication technologies were superseded by the context of capitalist political economies and the many institutions, organisations and other technologies that emerged to shape the biases of policy makers and publics. In one of his final essays, 'A Plea for Time', Innis recognised that interests with inordinate control over knowledge, wealth and force aspire to structure the mediators of contemporary culture in efforts to consolidate or extend control.

Mediators of this sort, for Innis, would include the American state through its ongoing efforts to control or expand the boundaries of US interests specifically, and capitalism more generally. Another is the institution of consumerism and its promotion of constant up-to-dateness and individualistic growth through commodities. It also, of course, involves commercial mass media whose efforts to extend and maintain ears and eyeballs compel them to promote the sensual here-and-now over relatively intellectual ruminations. More generally, it was the context of capitalism and its systemic tendency to dominate economic and other relations (not to mention the necessity, at least in its competitive form, to focus on short-term profits) that constitutes the context through which both the struggle for power takes place and the mediators of this struggle take shape. Under these conditions, as Menzies (2000) suggests:

> For all the contemporary talk about a postmodern information society, Innis's ideas would suggest that a real test of change is whether the social movements using the Internet . . . serve the bias of time – not just at the innovation stage and at the end-user level of intertextual rhetoric, but at the stage of institutionalized technological development and the enabling infrastructures associated with it, not just at the level of language games, but at the material level of structures that determine who gets to speak about what and who referees and designs the game plan.
>
> (Menzies 2000: 324)

An Innisian strategy

In pursuing this research focus, Innis's methodology directs us to assess Internet developments in the context of the world's predominant mode of production – capitalism. Since the late 1970s or so, structural changes have been taking place in the global political economy involving free trade and other neo-liberal policy reforms. As a result of associated, unprecedented and increasingly transnational fixed capital formations, the demand for technological developments designed to facilitate more efficiency has become extraordinarily important. One aspect of this systemic drive has been a dramatic extension in the capacity to profit from information-based products and services, sometimes generally referred to as the 'commoditisation of culture'.

Despite the significant and perhaps rising wave of non-commercial (and, sometimes, potentially counter-hegemonic) information and communication activities being accommodated by Internet-based developments, for the most part these new technologies are being developed and implemented to enable capitalist interests to expand their reach and improve efficiencies. All in all, the wealth and force under the direct or indirect control of the world's largest corporations and nation states constitute resources being used to promote the Internet as a means of increasing profits. In relation to this, what economist Ian Parker observed in 1988 remains insightful today:

> The commoditisation of culture has intensified the cultural differential between those individuals and institutions with financial resources to purchase, retreive and process large volumes of specialised and costly information and those who do not. At the same time, the increase in the average standard of living and leisure time and the extension of the mass media, particularly radio and television, have increased *general* access to a *basic* level of cultural programming which is literally unprecedented in global-historical terms. Particularly since 1945, we have thus witnessed the seemingly paradoxical phenomenon of a rapid and significant increase in the *absolute* general-informational density of advanced capitalist economies . . . combined with an increase in the *relative* concentration or *monopolisation* of specialised knowledge.
>
> (Parker 1988: 223–224, original emphases)

As a result of this historical context, and the role of Internet-related technologies in shaping it, knowledge is becoming an increasingly central means through which the production and reproduction of both capitalism and hegemonic order takes place. From a Marxist perspective, class rule requires the expropriation of material and non-material resources and a class's capacity to do this implies its relative control over key organisational resources. As in the past, the development of the Internet and related

technologies was a response to insecurity or crisis. Certainly a communication blackout following a Soviet nuclear attack (the basic incentive for
the precursor to the Internet's original funding by the US military) and,
later, what has been called the collapse of the post-1945 Fordist regime of
accumulation in the 1970s (Harvey 1990) would qualify as such moments
of insecurity and crisis.

The direct and indirect references to Marx in the preceding paragraphs,
indicating a number of similarities between Innisian and Marxist political
economy, serve to remind us that in the 1970s and 1980s several, mostly
Canadian, theorists argued that aspects of Innis's work, in fact,

> provide a means of dealing in dialectical materialist terms with several
> crucial lacunae in Marx's analysis: those of the dialectic between
> forces and relations of production and between the economic base
> and the superstructure; and at a more concrete level, those of the
> theory of the State and of the international economy that were to have
> occupied the unwritten fifth and sixth volumes of *Capital*.
>
> (Parker 1977: 548)

And while the debate over Innis's affinities or utilities for Marxist political
economy (and vice versa) has largely come and gone (Macpherson 1979;
McNally 1981; Parker 1983), the analytical and strategic possibilities of
relating the former's conservative and communication-focused dialectical
materialism with the latter's radical perspective remains pregnant with
possibilities (Comor 1994).

In relation to the contemporary post-Fordist period of rapid change
and, thus, insecurity, from an Innisian perspective, the Internet and many
other media both reflect such social-economic conditions and modify their
character. As Innis put it, 'the subject of communication offers possibilities
in that it occupies a crucial position in the organization and administration
of government and in turn of empires and of western civilization' (Innis
1986: 5). By 'civilization', Innis, of course, is referring to long-term
macro-structures and processes – a level of abstraction so removed
from here-and-now experiences that it was itself used by Innis as a frame
of temporal-spatial reference for both analytical and political-strategic
purposes.

An Innisian analysis

About fifty years after Innis's death, the Internet constitutes the most
significant of recent technological developments affecting how people

may relate to one another over time and space. As a medium of communication, it also reflects *and* restructures power relations (involving control over knowledge, wealth and force). Indeed, Innis's holistic understanding of communication media compels us to guard against any kind of Internet- (and even technology-) centred analysis. The Internet is just one of many significant mediators and communication scholars should be wary of assessing its development and implications in isolation from its historical context and the other technologies, organisations and institutions. Nevertheless, Innis's methodology directs us to think through the bias of Internet-based technologies in terms of the dialectics of power and control.

In its use in the annihilation of *both* time and space (at least in terms of the distribution and exchange of electronic forms of information) and in the context of the systemic pressure on capitalists, political leaders, workers and others to make decisions, buy commodities and take part in consumption activities more quickly and efficiently, the emerging bias of the Internet is disturbing indeed.

I use the word 'disturbing' for two reasons. First, the historically and technologically produced bias of the Internet to annihilate both time and space – its tendency to impel people to do much more in less time and with little regard for spatial barriers – challenges a broad range of vested interests and other communication media that tend to favour relatively long-term memory or decision making and/or various modes of spatial segmentation. Vested interests such as the labour movement or some domestically focused corporations, and media such as the book, or paper currency, or the nation-state, will no doubt continue to influence the temporal and spatial activities and orientations of people. As such, the Internet and related technologies constitute media that are, through their use, disturbing several established ways of doing and thinking.

The second reason for my choice of the word 'disturbing' to describe the bias of the Internet directs us to consider Innis's larger concern with how new communication media redress and/or stimulate other (or perhaps deeper) crises. In its implicit promotion of the short term – itself stimulated by the annihilation of spatial barriers such as nation-state borders (which could be used to 'buy time' for a culture, an economy or a government policy) – already we are experiencing disturbing trends. These involve the rapid erosion of the time to make decisions. Whether such decisions involve the bombing of an enemy, the security of one's investments, the options one has in the workplace and so forth, the Internet, the general commoditisation of culture and the related values placed on speed and efficiency arguably have set the stage for deepening political-economic crises as transnational investors respond to market 'signals' with spasmodic acts of panic selling, as consumers fail to keep up with the demands of sellers to buy more commodities more often, as the environmental crisis reaches a point of no return, and as cultures around the world become increasingly concerned with the here-and-now.

At a more personal level, efforts to promote the Internet and related technologies – through media ranging from the growing number of corporate interests promoting online consumerism, to educational organisations seeking access to vast information resources, to officials in World Bank development offices – explicitly or implicitly are serving the interests of those promoting the globalisation of capitalism. Beyond the efficiencies of instantaneous buying, selling, distribution and the agglomeration of information about consumer preferences, all of the world's people and places potentially will become networked in what Menzies calls a 'lego set of costlessly interchangeable production units' operating as a kind of transnational 'perpetual motion machine' (Menzies 2000: 331).

While an Innisian approach understands that media, once established and widely accessible, can be used in ways not intended by those initially structuring them, it does seem clear that if the main access points to the Internet continue to be dominated by profit-seeking interests and if corporate interests continue to be most influential in shaping national and international policies related to its development, it appears probable that the Internet will become, predominantly, a spatially biased medium. The more that this technology is structured to facilitate the ongoing growth and expansion interests of capital to the detriment of its potentials as an inclusive network in which vested interests concerned with conservation and duration (such as community groups, workers movements, religious organisations, environmental activists, and others) remain marginalised, existing controls over knowledge, wealth and force, for the most part, will be entrenched rather than challenged.

On the subject of knowledge, Innis was referring to not just what information is available and who has access to it but, more fundamentally, why we attend to the things to which we attend. In other words, in his elaboration of various media and their structured biases, Innis was concerned with the annihilation of civilisation coming when space or time becomes a cultural obsession. Through the Internet and predominant media, the practices and thoughts of more people in more parts of the world are becoming increasingly obsessed with immediate concerns and individual needs. Rather than a condition of capitalism *per se*, for Innisians, this is an accumulated result of the context of capitalism shaping and deepening the powers of some over others in conjunction with the spatially biased structures constructed to mediate day-to-day life.

In its moment-to-moment use, the Internet links many in relations directly or indirectly promoted by the systemic demand for efficiency. For many others, it links people in innumerable and instantaneous virtual communities. Either way, the relative intimacy of many non-commercial and face-to-face relationships tend to be pushed to the periphery of the human experience. At this juncture in history, the bias of the Internet is being structured and used in ways that diminishes time into the functionary of space.

Conclusion

For social scientists and, particularly, communication scholars, Innis's bias of communication constitutes an important analytical tool for three main reasons. First, bias directs us away from both technological and structural determinist positions precisely because its flexibility compels the analyst to recognise that, for the most part, physical or structural capacities at any given time and place are historically constructed. In Innis, such capacities are dialectically related to the intellectual and cultural capacities of human agents. As such, the bias of communication directs us toward a relatively sophisticated, critical, and materialist assessment of why we attend to the things to which we attend.

Second, bias enhances our ability to locate historic and contemporary sites of instability and crisis. Specifically, it directs us to consider the contradictory potentials of 'ways of thinking' and subsequent 'ways of doing'. Using bias as a conceptual tool, the seemingly successful short-term responses of vested interests to social-economic crises, for instance, can be seen to themselves entrench the very biases that contributed to the original crisis. As the case of environmental collapse illustrates, habits of socially structured thought are both historically produced and potentially disastrous.

Third, bias directs the researcher to pay particular attention to the core institutions, organisations and technologies used to mediate social-economic power relations. But again, because these media and their biases are socially constructed, the study of bias directs the social scientist away from reductionist and determinist modes of analysis. In light of this point, contemporary myths involving the Internet and related digital technologies could use a stiff dose of Innisian critical analysis. The fact that these history-shaping constructions have become 'inevitable' and 'desirable' compels the critical scholar to investigate the biases at play, the vested interests involved in their perpetuation, and the implications of struggles associated with them. This involves a focus on what institutions, organisations and technologies – what communication media – are most directly shaping such developments and their accompanying assumptions. According to Carey:

> What Innis recognized . . . is that knowledge is not simply informa-
> tion. Knowledge is not given in experience as data. There is no such
> thing as information about the world devoid of conceptual systems
> that create and define the world in the act of discovering it. And what
> he warned against was the monopoly of these conceptual systems or
> paradigms.
>
> (Carey 1975: 45)

In the context of the early twenty-first century, the Internet and other emerging technologies, organisations and institutions constitute the deep-ening predominance of an obsession with spatial expansion, organisation and control through ever-shortening time frames and an accompanying neglect of historical and social conceptualisations of time. As Innis put it in his essay 'A Plea for Time' in *The Bias of Communication*:

> a stable society is dependent on an appreciation of a proper balance
> between the concepts of space and time. We are concerned over
> control not only over vast areas of space but also over vast stretches
> of time. We must appraise civilization in relation to its territory
> and in relation to its duration. The character of the medium of
> communication tends to create a bias in civilization favourable to an
> over-emphasis on the time concept or on the space concept and only
> at rare intervals are these biases offset by the influence of another
> medium and stability achieved.
>
> (Innis 1982)

Half a century after his death, the concept of bias and Innis's dialectical materialist approach helps counter the guarded optimism held by some members of the intellectual movement that Collins, among others, has associated with him – postmodernism.

Today, Innis no doubt would be extraordinarily concerned with the trajectory of contemporary developments. While resistance would be anticipated, both the scale and rate of change associated with our obsession with growth, efficiency and immediacy – and thus the extreme difficulty of orchestrating sustained oppositional movements – would have surprised even him. The antidote to this state of affairs and its associated consol-idation of power through control over knowledge, wealth and force involves a concerted effort (perhaps, paradoxically, involving the Internet and other such technologies) to restructure existing and emerging means of mediating relationships involving the promotion of a collective critical memory and general sustainability. As anticipated by his dialectical triad, this effort to counter space with time will involve a near-future featuring tension rather than harmony. The alternative, for Innisians, most certainly involves a violent turn in the century now upon us.

Acknowledgements

Thank you to the editor, Christopher May, and to the reviewers of this chapter. Their critical and constructive comments have been very helpful. Thanks also to research assistant Erin Leonard.

Note

1 On its relation to the essays presented in *The Bias of Communication*, Innis writes that '[t]hey do not answer the question but are reflections stimulated by a consideration of it' (1982: xvii).

References

Acland, C.R. and Buxton, W.J. (eds) (2000) *Harold Innis in the New Century: Reflections and Refractions*, Montreal: McGill-Queen's University.
Carey, J.W. (1975) 'Canadian communication theory: extensions and interpretations of Harold Innis' in G. Robinson and D. Theall (eds) *Studies in Canadian Communications*, Montreal: McGill University.
Collins, R. (1986) 'The metaphor of dependency and Canadian communications', *Canadian Journal of Communication* 12: 1–19.
Collins, R. (1989) 'Book review – *Empire and Communications*', *Philosophy of the Social Sciences* 19: 217–219.
Comor, E. (1994) 'Harold Innis' dialectical triad', *Journal of Canadian Studies* 29: 111–127.
Couch, C.J. (1990) 'Mass communication and state structures', *Social Science Journal* 27: 111–128.
Cox, R.W. (1995) 'Civilizations: encounters and transformations', *Studies in Political Economy* 47: 7–31.
Harvey, D. (1990) *The Condition of Postmodernity*, Oxford: Blackwell.
Innis, H.A. (1935) 'The role of intelligence: some further notes', *Canadian Journal of Economics and Political Science* 1: 280–287.
Innis, H.A. (1940) *The Cod Fisheries*, Toronto: University of Toronto.
Innis, H.A. (1946) *Political Economy in the Modern State*, Toronto: Ryerson Press.

Innis, H.A. (1982) [1951] *The Bias of Communication*, Toronto: University of Toronto Press.

Innis, H.A. (1986) [1950] *Empire and Communications*, Victoria: Press Porcepic.

Innis, H.A. (1995) *Staples, Markets, and Cultural Change*, Montreal: McGill-Queen's University.

McNally, D. (1981) 'Staple theory as commodity fetishism: Marx, Innis and Canadian political economy', *Studies in Political Economy* 6: 35–64.

MacPherson, C.B. (1979) 'By Innis out of Marx: the revival of Canadian political economy', *Canadian Journal of Political and Social Theory* 3: 134–138.

Marvin, C. (1983) 'Space, time, and captive communications history' in M.S. Mander (ed.) *Communications in Transition*, New York: Praeger.

Menzies, H. (2000) 'The bias of space revisited' in C.R. Acland and W.J. Buxton (eds) *Harold Innis in the New Century: Reflections and Refractions*, Montreal: McGill-Queen's University.

Parker, I. (1977) 'Harold Innis, Karl Marx and Canadian political economy', *Queen's Quarterly* 84: 545–563.

Parker, I. (1983) '"Commodity fetishism" and "vulgar Marxism": on "Rethinking Canadian political economy"', *Studies in Political Economy* 10: 143–172.

Parker, I. (1988) 'Economic dimensions of 21st-century Canadian cultural strategy' in I. Parker, J. Hutcheson and P. Crawley (eds) *The Strategy of Canadian Culture in the 21st Century*, Toronto: TopCat.

Urwick, E.J. (1935) 'The role of intelligence in the social process', *Canadian Journal of Economics and Political Science* 1: 64–76.

Chapter 6

Lewis Mumford

Christopher May

> The inventors of . . . computers are the pyramid builders of our own
> age: psychologically inflated by a similar myth of unqualified power,
> boasting through their science of their increasing omnipotence, if not
> omniscience, moved by obsessions and compulsions no less irrational
> than those of earlier absolute systems: particularly the notion that
> system itself must be expanded, at whatever eventual cost to life.
>
> (Mumford 1964: 5)

Too often the history of ICTs is wrenched out of the history of technology
and presented as something altogether separate and therefore different,
rendering previous analyses irrelevant. However, Lewis Mumford, who
died before the information age was widely proclaimed, would have found
little novel about the general dynamic of development of information
and communication technologies. Indeed he would have found the debates
about the role of new technologies in our lives very familiar. In this chapter
I recover Mumford's ideas about technological development and relate
them to our 'new age'. After a brief overview of Mumford's work, I
discuss some of his ideas regarding technology and suggest their relevance
to the contemporary emergence of the information society. I conclude
that our understanding of the contemporary technological 'revolution' can
be considerably enhanced by looking back to Mumford's conclusions
about the much longer term history and development of technology in
society.

Who is Lewis Mumford?

Lewis Mumford was born in 1895 in Flushing, New York and died at
home in Amenia, New York aged 95, having written over thirty books,
countless articles and reviews. Mumford spent his youth in New York,

walking around the city, watching construction and exploring the growing metropolis. Indeed as Goist points out, 'the city was the very stuff of his life' well before he became seriously interested in studying cities (Goist 1972: 380). But the young Mumford was also interested in technology. His first published article, aged 13 and prompting a royalty cheque for 25 cents, was in *Modern Electrics* on new breakthroughs in radio receivers, reflecting his own interest in building and experimenting with radios (Miller 1989: 34). After the end of the First World War, Mumford studied with Thorstein Veblen at the New School of Social Research, who became both a friend and one of his major influences. He also discovered the work of Patrick Geddes (the Scottish botanist and social philosopher) with whom he corresponded until Geddes' death in 1931. It was Geddes' influence that led Mumford to think about the role of cities in the history of humankind.

Mumford was interested in the human possibilities that sprang from humankind's organisation into cities, possibilities which he observed were often compromised. Therefore he sought to establish a more humane approach to the development and planning of cities. To this end for some years he wrote a column in the *New Yorker* which aimed to widen the debate on architecture and planning (specifically in New York, but also more generally). For Mumford, the key problem, revealed by humankind's organisation into cities, was the impact of technology. Increasingly he studied the historical role of technology, and explored its impact on social relations (and their impact on technology). While Mumford was not so much a historian as a synthesiser of disparate ideas and analyses, Jamison has claimed that his book *Technics and Civilisation*

> created a new field of study: history of technology . . . Mumford succeeded in placing technological development in human context; after Technics and Civilisation, the debate about technology moved to a new level of constructive ambition and seriousness. It would never be quite the same again.
>
> (Jamison 1995)

Especially after the 1950s, Mumford's focus shifted almost completely to the issue of humans' relation with technology, and the problem that technology seemed to be running 'out of control'. Mumford was not in any way a technological determinist, but despite some clear parallels between Mumford and the social-embeddedness-of-technology approach, he did not merely hold an earlier version of this paradigm. Indeed Lewis Mumford is a singular character and part of no 'school'. Though he did

experience the earlier stages of the information 'revolution', he died on the
eve of the information technological acceleration of the 1990s. Thus,
the interpretation I develop below applies Mumford's ideas to a technological history he did not fully experience. But, the central argument of this
chapter is that his approach remains relevant to this recent history.

Mumford's perspective on the history of technology was informed
by his interests and political views, but as importantly by his emphasis
on 'symbolic activities'. He resisted the commonplace material analysis
of technological advances, an account linking a series of material technologies with little regard for their symbolic values or importance. Indeed, for
Mumford the

> constant danger in interpreting human behaviour is to overvalue exact
> methods and measurable data, separated from their historical context:
> data often too complex for even verbal formulation; for the very things
> that the conscientious historian is tempted to leave out, because of
> their obscurity, their purely analogical suggestiveness, their subjective
> involvement, are needed to bring any richness into our judgements.
>
> (Mumford 1962: 202)

This is encapsulated in his dialectic of materialisation and etherialisation,
the interaction of material or tangible elements of life with the subjective
forces of abstract ideas or symbolic representations.[1] Like Hegel, Mumford
saw historical progress as being the process of actualisation of symbolic
and ideational or abstract ideas. But, and here the difference with Hegel
is important, once such actualisation had been accomplished, the new
material artefacts might themselves prompt an ideational change in a
continuing (and continual) process. History was driven by a bi-directional,
dialectic process of interaction between the material and the symbolic
(Mumford 1971: 421–429). Therefore the material artefacts which could
be uncovered and investigated in the quest for a history of technology
immediately skewered such a history towards an overemphasis of the
material aspects of technology relative to the ideational aspects.

Human beings' relation with technology is not as passive receivers of
innovation but as shapers of the cultural drivers of technological advance;
human ideas and symbolic concerns are major factors in the history of
technology. Mumford therefore stressed the human agency in the history
of technology. As early as 1924 he argued that the

> future of our civilisation depends upon our ability to select and control
> our heritage from the past, to alter our present attitudes and habits, and

to project fresh forms into which our energies may be freely poured
. . . During the last century our situation has changed from that of
creators of machinery to that of creatures of the machine system; it
is perhaps time that we contrived new elements which will alter once
more the profounder contours of our civilisation.

(Mumford 1924: 195–196)

And ten years later, in the preface to *Technics and Civilisation* he warned
that

no matter how completely technics relies upon the objectives procedures of the sciences, it does not form an independent system, like the
universe: it exists as an element in human culture and it promises well
or ill as the social groups that exploit it promise well or ill . . . In order
to reconquer the machine and subdue it to human purposes, one must
first understand it and assimilate it. So far, we have embraced the
machine without fully understanding it, or like the weaker romantics,
we have rejected the machine without first seeing how much we could
intelligently assimilate.

(Mumford 1934: 6)

This danger of being controlled by technology rather than shaping it
ourselves is no less evident now than it was then.

Thus, Mumford's approach 'has the great merit of treating technology
as a system of social relations and of recognising the *mutual* dependence of
technological change and social change' (Lasch 1980: 24, emphasis added).
Recognising his own part in society's contemporary dialectic, the critique
of the uses of contemporary innovations is prominent in Mumford's discussion of the history of technology. But he 'never forgot that modern culture
has itself given rise to the critical traditions that most adequately explain the
contemporary crisis and point the way to its resolution' (Lasch 1980: 28).
Crucially, Mumford was not anti-modern, nor a cultural conservative, but
rather believed in the ability of humankind to reshape the future despite the
claims for the inevitability of the effects of technological progress.

Like Ruskin before him, Mumford regretted that technology, alongside
the division of labour, has 'separated intellectual and practical life, sacrificed values to technique, and given rise to a deadening professionalism'
(Casillo 1992: 96). This led Mumford to call for a return to craft, not
in the sense of a stand against technology, but rather in the interests of
serving human needs and 'organic' forms of community. Langdon Winner
suggests that Mumford saw technology as a potentially liberating force,

'by which humanity will free itself from the snares of militarism, greed, power worship, lopsided epistemology and exploitative attitudes towards nature and fellow humans' (quoted in Winters 1987: 107). Though technology is the problem it is also the potential solution. For Mumford, technology is not an uncontrollable force, but rather is controlled by social forces: thus the favouring of particular practices and uses is not only possible and practicable but is how technology has always been deployed and used. But there is some ambiguity here:

> it is often unclear whether [Mumford] believe[s] industrial technology can be reconstructed to achieve [his] goals, or whether to reject it in favour of simpler craft technology. Does the social determination of technology concern alternatives *within* industrialism or merely the choice *between* industrial and craft technology?
>
> (Feenberg 1991: 125; original emphases)[2]

Certainly, while I lean towards an appreciation of Mumford's perspective which suggests there is a choice *within* industrialism, it is difficult if not impossible to dissolve this ambiguity within his work. But despite this ambiguity Mumford's analysis remains apposite to contemporary discussion of the information society.

Mumford also regarded the notion of 'economic man' as absurd. It was 'only in economics textbooks . . . the Economic Man and the Machine Age have ever maintained the purity of their ideal images' (Mumford 1934: 269). Having studied with Thorstein Veblen, Mumford's economic perspective was influenced by Veblen's view that the ruling class are a 'tumour' on the body of society, which

> has resulted in the general devaluation of productive work in favour of conspicuous waste and leisure. Confusing money values with real wealth, capitalism sacrifices production and the instinct of workmanship to the pecuniary interests of businessmen and financiers.
>
> (Casillo 1992: 102)

But unlike Veblen, Mumford was concerned not only with productivity and the powers of the technocracy, but also with the ability of human beings to benefit from certain technological developments. This led Mumford to an understanding of the history of technology that commenced not with the first machine, or even the first tool, but with language, and the desire to modify the body in relation to its environment (dress, heat, physical fitness for hunting).

Technology commences with people's own mental activities directed toward changing things. Indeed as Mumford points out, while technological history has been fixated on the fashioning of tools,

> the invention of language – a culmination of man's more elementary forms of expressing and transmitting meaning – was incomparably more important to further human development than the chipping of a mountain of hand-axes . . . For only when knowledge and practice could be stored in symbolic forms and passed on by word of mouth from generation to generation was it possible to keep each fresh cultural acquisition from dissolving with the passing moment or the dying of a generation. Then, and then only, did the domestication of plants and animals become possible.
>
> (Mumford 1966a: 308)

The effort to develop the ability to speak, to develop this technology of communication, enabled the dissemination and storage of experience which could then be improved and built on. Only then could material technologies be developed through the collectivisation of experience and the organisation of effort.

This reintegration of the ideational leads Mumford to stress throughout his work that neither human culture nor technology can be understood without the other. But it must be stressed that Lewis Mumford was not anti-technology, as Arthur Molella noted, he actually 'savoured technological processes for their own sake as unique manifestations of human ingenuity' (quoted in Morley 1989: 123). Nevertheless,

> although he acknowledges the importance of material conditions in society and culture, Mumford asserts the relative autonomy of man's 'idolum' or Weltanschauung and so rejects as inorganic (hence mechanistic) the vulgar Marxist view that ideas, values and aesthetic symbols merely reflect or conceal material factors.
>
> (Casillo 1992: 92)

This organicism is a major element in the way Mumford views the relations between humans and machine. In one sense Mumford's organicism is a Darwinism modified by the influence of Peter Kropotkin (Casillo 1991: 95; Ward 1986: 13–14). Though familiar with the work of Karl Marx through his contacts with New York socialist groups in the years before the First World War, Mumford himself did not adopt a Marxian perspective on technology. As Miller points out, 'his desire for a more humane economic system emerged from Plato, Ruskin, Morris, Tolstoy and Kropotkin, not from Marx and Engels' (Miller 1989: 99). In Mumford's view

> Kropotkin realised that the new means of rapid transit and commu-
> nication, coupled with the transmission of electric power in a network,
> rather than a one-dimensional line, made the small community on a
> par in essential technical facilities with the over congested city. By the
> same token, rural occupations once isolated and below the economic
> and cultural level of the city could have the advantages of scientific
> intelligence, group organisation and animated activities, originally a
> big city monopoly; and with this the hard and fast division between
> urban and rural, between industrial worker and farm worker, would
> break down too . . . With the small unit as a basis, he saw the oppor-
> tunity for a more responsible and responsive local life, with greater
> scope for the human agents who were neglected and frustrated by
> mass organisations.
>
> (Mumford 1975 [1961]: 585–586)

In his interpretation of Kropotkin's *Mutual Aid* the relevance of Mumford's view of technology for the global information society is clear, if the city's advantages are further extended through the Internet and ICTs. The widening of accessibility to the advantages of the city allow the more organic, smaller human communities which Mumford regards as being more 'democratic' to develop similar human advances as have been bought at too high a price in the city.

Despite his appeal to 'democratic' technics, Lewis Mumford has not infrequently been criticised for a form of right-wing, implicitly conser-vative, communitarianism. This is partly a result of his pre-1940s work on cities and planning which exhibited a preference for technocratic, centralised bureaucratic control (not an uncommon view in the 1930s). However, by the 1960s, when his interest had turned almost exclusively to the problem of technology, he rejected the sorts of utopian visions which had influenced him in the pre-war period, seeing them as essentially totalitarian. In any case his utopian vision had always been somewhat compromised: even in his *The Story of Utopias* (written in 1922), ironically 'Mumford finds utopias wanting because they are out of contact with the everyday world of real men and women' (Goist 1972: 383). In later life he retained his organicism, but linked it to a devaluation of the role of machinery in the health and development of the organic community (Casillo 1992: 113). However, the advance of technology may constrict the available choices open to the individual, and thus deny the possibility of reinvigorating the individual's links with nature. The human environment is increasingly, for the older Mumford, a technologically constructed megalopolis. Thus he saw

> the task of a cultural critic was not merely to identify problems, but to provide constructive ideas for bringing the machine under human control, and even more, using machinery to enrich human life. It had become clear [even in the early 1930s] that the machine needed a conscious programme for its guidance; and it was this that Mumford aimed to provide, by so doing pioneering the social study – and constructive assessment – of science and technology.
>
> (Jamison 1995)

And it is this stream of Mumford's thought, the constructive engagement with contemporary technology through the application of an understanding of its complex history, that I explore next.

Lewis Mumford and technics

Lewis Mumford preferred to discuss 'technics' rather than technology. This term encapsulates for Mumford the importance of human interaction with technology: technologies should be located within their social relations, their context, before assessing their social value and effect. Mumford suggested that these technics could be broadly regarded as either authoritarian or democratic.[3] This distinction does not necessarily map onto *specific* technologies, but historically the usage and development of technology fits into one or other of these tendencies. At the centre of his treatment of technics is the recognition that both

> have recurrently existed side by side: one authoritarian, the other democratic, the first system-centred, immensely powerful, but inherently unstable, the other man-centred, relatively weak, but resourceful and durable.
>
> (Mumford 1964: 2)

In this dialectic, centralising authority tries to control the use and outputs of technology, but is unable to completely micro-manage the society in which it exists. A space for resistance remains available, where technologies may emancipate and empower individuals *against* the authoritarian power. It is this possibility that supports and reproduces democratic technics.

Mumford's understanding of democracy is centred on the interaction of small groups, allowing 'communal self government, free communication as between equals [and] unimpeded access to the common store of knowledge' (Mumford 1964: 1). Institutional arrangements which locate authority at the apex of an organisation, and centralised direction of activities, are therefore seen as threats. However, he accepted that large-scale societal organisation was required to produce some major social benefits. But to forestall the move to complete authoritarianism the whole system should be cut back to 'a point at which it will permit human alternatives, human interventions, and human destinations for entirely different purposes than those of the system itself' (Mumford 1964: 8). There is a distinction between 'labour-saving' technologies that free individuals for more fruitful tasks, and technologies that merely reinforce (or even expand) 'drudgery, enlisting human energy in collective enterprises on a gigantic scale . . . that gratify the power-hunger of the mighty but do little to improve the material conditions of everyday life' (Lasch 1980: 23). Authoritarian technics in the nineteenth and twentieth centuries have enabled the logic of control through the 'mega-machine' to move beyond mass-enterprises and enter everyday working lives. This, argued Mumford, was the true 'industrial revolution', not the introduction of mechanisation and automation but rather the division between planning and production, between conception and execution. The industrial revolution was in essence a revolution in the social relations of knowledge. Thus, if democratic technics can be reasserted, instead of

> liberation *from* work being the chief contribution of mechanisation and automation . . . liberation *for* work – for educative, mind-forming work, self-rewarding even on the lowest psychological level – may become the most salutary contribution of a life centred technology.
> (Mumford 1966a: 316, original emphases)

But such a desire is compromised by the continuing dominance of authoritarian technics.

Authoritarian technics

For Mumford, authoritarian technics first emerged during the period of pyramid building in Egypt. Collecting together vast mega-machines (of organic components – men, women and children) to do their bidding, utilising the new skills of communication/writing, mathematics and bureaucratic or organised control, the God Kings constructed structures that were beyond the capabilities of previous technics. In a sense, the ability to organise society to specific ends marks the dawn of 'civilisation' in Mumford's eyes, even if such civilisation brings with it the problem of authority and domination. To build the pyramids, technics of this scale relied on centralised political control, 'ruthless physical coercion, forced labour and slavery' to bring into existence 'machines that were capable of exerting thousands of horsepower centuries before horses were harnessed or wheels invented' (Mumford 1964: 3). The unleashing of such potential came at a high psychological and physical cost to most of the individuals involved. For Mumford this is the recurring problem in the history of technology.

Authoritarian technics met little resistance in the first instance because they produced an 'economy of controlled abundance', allowing a vast expansion in the efficiency of agriculture, and the ability to support large urban populations, as well as releasing an increasingly influential elite of bureaucrats, scientists, religious and military functionaries from physical labour of any sort. The first wave of authoritarian technics (reaching its apogee for Mumford, with the Roman Empire) could support the emergence of new technologies only in urban centres. Mumford argues that these first authoritarian technics finally proved too 'irrational' to continue indefinitely, dependent on the centre retaining control: once communication failed and authority was no longer regarded as legitimate, the mega-machine collapsed. And since

> authoritarian technics first took form in an age when metals were scarce and human raw material, captured in war, was easily convertible into machines, its directors never bothered to invent inorganic substitutes.
>
> (Mumford 1964: 4)

Thus, as the supply of slave and forced labour declined with the absorption rather than defeat of populations and the problems of controlling a vast empire, so the logic of authoritarian technics and its social reality drifted apart. This allowed a more democratic technics to reassert itself during the Middle Ages, only to be once again constrained by the rise of the nation-state, the mega-machine *par excellence*.

For Mumford the Enlightenment and the scientific revolution led to a view that technological development and scientific progress would produce an increasingly democratic society.

> But what we have interpreted as the new freedom now turns out to be a much more sophisticated version of the old slavery: for the rise

of political democracy during the last few centuries has been increasingly nullified by the successful resurrection of a centralised authoritarian technics . . . At the very moment Western nations threw off the ancient regime of absolute government, operating under a once-divine king, they were restoring this same system in a far more effective form in their technology, reintroducing coercions of a military character no less strict in the organisation of a factory than in that of the new drilled uniformed army.

(Mumford 1964: 4)

The powerful have constructed a system in which their power over technology underlines their claims for omnipotence. Thus, Mumford argues, in an age of authoritarian technology there is no longer a 'visible personality, an all-powerful king' a sovereign location of power, rather it is the system itself that is now authority (Nitzan 1998: 204). And the perception of the system as providing the limits to action (and possibility) rather than an actual (locatable) ruler, helps authority defuse most of the resistance from democratic technics. This is not to argue there are not individuals or groups with power in society, but that such power is partly masked by the technological system's 'needs'.

In a striking precursor to critical writing regarding the globalisation of liberalism, and its structuring of information society, Mumford asserts the authority of technics is defined through its logic, the promotion of efficiency:

Under the pretext of saving labour, the ultimate end of this technics is to displace life, or rather, to transfer the attributes of life to the machine and the mechanical collective, allowing only so much of the organism to remain as may be controlled and manipulated.

(Mumford 1964: 6)

What allows this system to reproduce its position is the ability to provide for the majority an abundance of material goods without historical precedent. However, this is possible only where non-systemic wants are not articulated, where only deliverable demands are acceptable. And it is here that a crack in the authoritarian facade opens. For Mumford it is the historical process of self-discovery, the ability of humans to change, that always undermines the ability of authoritarian technics to retain control without constant (and contested) reproduction.

Democratic technics

Mumford stresses such overarching authoritarian technics are not necessarily a reality but a possibility, one which can be guarded against only by valuing and supporting democratic technics. Furthermore the bribe of abundance and material wealth which authoritarian technics offers in return for a narrowing of human potential and a decline in psychological health can be rejected. If democracy itself is to be supported and guarded, then technology must not be seen only as a tool which automatically brings with it empowerment or enslavement. Technological deployment and effects reflect the social relations in which they appear, and perhaps most importantly *no* technology is beyond systemic incorporation into authoritarian technics. It is imperative for Mumford that the human scale of life be central to democracy, society must revolve around humans not the system (Mumford 1964: 8). Thus, technology, in itself neither authoritarian nor democratic, must be *positively* integrated into a democratic technics.

In contrast to authoritarian technics, democratic technics are small scale and 'even when employing machines, remain under the active direction of the craftsman', responding to their needs and wants (Mumford 1964: 3). Human-scale technologies have modest demands (which is to say localised power needs, locally available skills, low organisational requirements) and can be adapted to local conditions. Despite the authoritarian technics of contemporary society, there remains the potential for localised and democratic technics to retain a level of autonomy, and thus the ability for local creativity to be exercised. Mumford forcefully argues that society needs to look beyond abundance of material goods, and examine the needs of individuals. This will require a shift in

> the seat of authority from the mechanical collective to the human personality and the autonomous group, favouring variety and ecological complexity, instead of stressing uniformity and standardisation, above all, reducing the insensate drive to extend the system itself, instead of containing it within definite human limits and thus releasing man himself for other purposes.
>
> (Mumford 1964: 8)

Democratic technics free the individual from the burden of continual employment, allowing the development of non-system-oriented behaviour. Mumford proposes the emancipation of the creative individual: democratic technics allow work that is dependent upon 'special skill, knowledge, aesthetic sense'. Large-scale enterprise may continue but there needs to be a space for individual expression through artisanal activities within the localised community.

Mumford also argues that in the past democratic technics have supported a situation where

> the only occupation that demanded a life-time's attention was that of becoming a full human being, able to perform his biological role and to take his share in the social life of the community . . . Every part of work was life work.
>
> (Mumford 1966b: 238)

The development of self can continue under democratic technics, even if selfhood is stifled by the demands of authoritarian technics. Indeed the notion that democratic technics should abolish all work is far from Mumford's mind: 'work which is not confined to the muscles, but incorporates all of the functions of the mind, is not a curse but a blessing' (Mumford 1966b: 242). Though democratic technics potentially exist they need to be positively constructed. Thus, if 'the fashionable technocratic prescriptions for extending the present system of control to the whole organic world are not acceptable to rational men, they need not be accepted' (Mumford 1971: 430). Resistance to the mega-machine, against authoritarian technics, is possible.

Indeed, as Mumford suggests:

> Nothing could be more damaging to the myth of the machine, and to the dehumanised social order it has brought into existence, than a steady withdrawal of interest, a slowing down of tempo, a stoppage of senseless routines and mindless acts.
>
> (Mumford 1971: 433)

And therefore, most importantly for the debates regarding the information society, authoritarian and democratic technics are not defined by their technologies but by their use. Thus, technologies do not have a natural character, they do not automatically support or destroy democracy, but rather help reproduce social structures and systems. Consequently, democratic technics and authoritarian technics do not replace one another, but rather exist side-by-side, in competition, ebbing and flowing but never finally erased. Withdrawal from areas of the economy such as the minority but symbolic 'downshifting' within the middle class, and the resistance to capitalism utilising its own technologies, such as recent 'reclaim the streets' protests organised through chat-rooms and newsgroups on the 'Net', seem to fit Mumford's notion of the reassertion of democratic technics quite well.

Megalopolis and information society

In the inter-war period Lewis Mumford was best known as an advocate of regional planning and its impact on the well-being of city dwellers. One aspect of Mumford's work on cities is of some relevance to the emergence of an information society: the problems which arise from the spreading megalopolis. From village, to the polis, to the metropolis, Mumford saw urban history as a centrifugal force enlarging the city. The Megalopolis is the final stage of urban development before the collapse into necropolis, the hollowed-out city of the dead.[4] Like the information society, the mega-lopolis has wrenched itself free of its material surroundings to encompass a growing territory through its communication networks. While the centre holds, the metropolis's control and influence is progressively widened by communication. In the information society the megalopolis has transcended geographical limitations to become the global city.

At the centre of Mumford's positive view of the city was the enhanced possibility of varied and constructive human interaction. Like writers before him he also valued the freedom from traditional social hierarchies the relative anonymity the city offered. The city functions not only as an information exchange between regions (individuals are drawn to the centre to interact), but also as a social magnet which attracts and spurs innovations in social and cultural practices which benefit from, and spread through, the constant social interactions the city engenders by the proximity of different groups and functions. As importantly, the access to a large population enables specialisation and more selective service provision. The city allows a division of labour which can enhance the life of its inhabitants by making new services and products available, ones which could never have been supported by smaller settlements. However, under an authoritarian technics the division of labour also leads to the progressive deskilling and intellectual impoverishment of the workforce, unless it is balanced by the retention of democratic technics in one form or another.

This division within the megalopolis is also one between 'town and county' inasmuch as Mumford sees the rural village, the non-metropolitan communities, as reservoirs of renewal from which the centre can draw new talent, and humanity in times of need, or after the destruction (collapse) of the metropolis (Mumford 1975 [1961]: 636). However, as cities grow larger these positive elements of their character become compromised until

> instead of producing the maximum amount of freedom and spon-
> taneity, this scattering of the metropolitan population over the remoter
> parts of the countryside confines its working members for ever-longer
> periods to a mobile cell, travelling ever-longer distances to the place
> of work or to achieve even a few of the social and interpersonal
> relations that the city once provided at one's elbow.
>
> (Mumford 1968 [1962]: 131)

The movement of the population into commuter towns and suburbs undermines the value of the crowded city, while reinforcing its tendency to alienation and atomisation. The city was always a mixed blessing but the megalopolis retains the problems while dissipating the advantages.

In a key passage, Mumford argues that the megalopolis (as anti-city)

> combines two contradictory and almost irreconcilable aspects of
> modern civilisation: an expanding economy that calls for the constant
> employment of the machine (motorcar, radio, television, telephone,
> automated factory, and assembly line) to secure both full production
> and a minimal counterfeit of normal social life; and as a necessary
> offset to these demands, an effort to escape from over regulated
> routines, the impoverished personal choices, the monotonous pros-
> pects of this regime by daily withdrawal to a private rural asylum,
> where bureaucratic compulsions give way to exurban relaxation and
> permissiveness, in a purely family environment as much unlike the
> metropolis as possible. Thus the anti-city produces an illusory image
> of freedom at the very moment all the screws of organisation are being
> tightened.
>
> (Mumford 1968 [1962]: 132)

The megalopolis allows the further intensification of 'productive' activities in the metropolis by providing a form of respite from its pressures at the centre. The megalopolis allows the diffusion of the functions of the centre while retaining the metropolitan (which is to say centralised) control of these functions. Additionally, the arrival of instantaneous communication has allowed (and encouraged) further concentration of bureaucratic power. The centre not only controls and directs distant economic activities but also influences and recasts distant cultures and practices in its image (Mumford 1975 [1961]: 608). In this megalopolis the distinction between the city and its immediate environs disappears, there is expansion at the edges and previously separate settlements, towns or in extreme cases neighbouring cities are swallowed up. The megalopolis swallows all around it and merges with its surroundings, no longer clearly discernible but slowly and surely destroying diversity. More importantly for my purposes here, the megalopolis in its ethereal form is the information society itself.

Though the megalopolis may be hellish, it may also if allied to demo-cratic technics allow the advantages of the city to be enjoyed more widely. Mumford suggests that communication technologies and the organisation of social activities in the megalopolis may produce an 'invisible city' (Mumford 1975 [1961]: 641ff). The emergence of this communication mediated invisible city may allow a further division of labour between it and the visible city, allowing a space for relations that thrive on physical proximity in the visible city, while services and practices previously only maintained in the visible city can be spread through the invisible city. The functions of the city, previously limited to the metropolitan centre are distributed through a 'functional grid', the framework of the invisible city. Positively, this grid and its associated networks

> permit units of different size, not merely to participate, but to offer their unique advantages to the whole . . . [one location] can be an effective part of the whole, making demands, communicating desires, influencing decisions without being swallowed up by the bigger organisation.
> (Mumford 1975 [1961]: 644)

However, 'the new grid, in all its forms, industrial, cultural, urban, lends itself to both good and bad uses' (Mumford 1975 [1961]: 642). And the negative elements of the megalopolis or invisible city are hardly trivial.

Writing just before the Second World War, Mumford had argued that conflicts would result from the centralisation of economic organisation in the megalopolis, a centralisation with little regard for territorial borders. And such conflicts would lead ultimately to war, or at the least a vast military-industrial complex (to use Eisenhower's term) which continually used the threat of instability to gather resources to itself (Mumford 1940: 272–273). Indeed, Eisenhower's farewell address, from where the term derives, itself can be read as a warning of the need for vigilance in the face of an authoritarian technic.[5] Additionally, within the megalopolis everything is subject to the market, including such cultural activities as education and the arts. In this sense, the megalopolis represents the inten-sification of capitalism, which as I have argued elsewhere (May 1998) is one of the key elements of the information society.

While the metropolitan centre retains the organs of culture concentrated in the previous stage, and as such sees less favourable developments tempered, in the vast sprawl of the megalopolis the increasing domination through the functional grid can lead to an impoverishment of existence. In the megalopolis

> secret knowledge has put an end to effective criticism and democratic control; the emancipation from manual labour has brought about a new kind of enslavement: abject dependence upon the machine. The monstrous gods of the ancient world have all reappeared, hugely magnified, demanding total human sacrifice . . . [and] whole nations stand ready, supinely, to throw their children into [their] fiery furnace.
> (Mumford 1975 [1961]: 651–652)

But despite this possibility, Mumford's reading of technological history helped him locate a possibility of deliverance from this situation. The development of a democratic technics within the functional grid, to balance the centralising tendency of the authoritarian technics at the centre of the megalopolis, could redress this imbalance and allow the retention of humanity and a healthy variance in culture across the system.

The megalopolis in its form of the invisible city is the clearest precursor to the information society within Mumford's work. Though the terminology is different there is a clear correlation with the opportunities and problems of the information society. The information society, it is supposed, allows access from anywhere (or at least anywhere within the network) to the cultural, informational and political assets that are distributed throughout the system. The information society as megalopolis has no formal centre but enables immediate contact with any of its component parts (recall the origins of the Internet in the amorphous command and control structure of the US military's ARPANET, designed to have no fixed structure of communication, to allow information to flow even if elements where 'knocked out'). As with the megalopolis, the information society is both empowering and limiting of human experience. Though intellectual resources are now available, there is also a centralisation of power, an increased capacity for control and surveillance. The metropole retains its power while progressively swallowing up other areas of human activity and giving an impression of individual empowerment.

Lewis Mumford and the twin dynamics of the information society

Lewis Mumford was neither the first (nor the last) to recognise the socially constructed nature of technology usage. There has been a long and distinguished history of criticism of the role of technology in the impoverishment of psychological existence, perhaps best typified by the tradition of Luddism (Robins and Webster 1999: 39–62). Mumford presents these criticisms within an account of technological development that proposes a recurring pattern, not isolated moments of upheaval. And thus, while Mumford died on the eve of the acceleration of the information revolution, his analysis of competing (and coexisting) technics is still useful for thinking about the global information society. Furthermore, his view of technics as not being exclusively manifest in material artefacts fits well with information society's concentration on the use and deployment of knowledge and information. For Mumford, the use of knowledge, even in the complex manner claimed for the new era is actually a recurrent and continuing factor in technological history.

Mumford's analysis supports the recognition of the continuity of technological practices in the information society, and a continuity with the expansion of the city. It allows the identification of a spatial continuity between the process of city-building and the emergence of a virtual political economic space; the 'invisible city' existing electronically. Mumford's view of technological advance firmly locates the information society in the ongoing history of technics. In this sense Mumford's (implicit) view on the information society stresses the continuity with previous organisational moves within capitalism, and as such shows a similarity with a more critical stream of work on the recent history of information society (Robins and Webster 1999; Beniger 1986). This perspective on the information society looks back at the 'second industrial revolution', the emergence of mass-production, scientific management and other aspects of modern capitalism to stress recent developments' link with previous 'advances' in both manufacturing and services.

Lewis Mumford's conception of the history of technology suggests the interconnected nature of authoritarian and democratic technics. The balance between them might shift depending on particular social relations but neither was the only tendency within technological change. This insight seems to me to be directly relevant to debates around the emergence of the information society. Thus while the terms may have changed, the aspects of technological development that Mumford used to construct his idea of technics have not disappeared. The notion of authoritarian and democratic technics runs parallel to the distinction that divides opinion regarding the information society: whether its dynamic is 'disclosing' or 'enclosing'. On each side of this argument one dynamic is regarded as normal while the other is regarded as a temporary aberration which will wither as the information society continues to develop. This bifurcation of normal/abnormal results in the reproduction of partial perspectives on the information society. Utilising Mumford's work, such a separation can be dissolved, but first let me summarise the current, and widespread division of opinion over the issue of information society's underlying dynamic.

Authoritarian/enclosing tendencies

One broad group of commentators focus on an enclosing dynamic and regard the information revolution (the technological backdrop to information society) as involving an intensification of property relations. The ability to render knowledge and information as intellectual property rights (IPRs) suggests the information society represents an expansion of modern capitalism, not its replacement (Bettig 1997; Kundnani 1998). This perspective criticises claims for a new period of social organisation made on the basis that the raw materials which are the subject of economic activity (along with the organisational structures of such activities) having been profoundly altered by the rise and expansion of informational economics. But, the ability of economic actors to treat new forms of products and services as (intellectual) property suggests continuity not disjuncture. The underlying character of capitalism has always been the relation between property holders and those who only have their labour to bring to the market (May 1998). The information society is merely business as usual. These critics of information society have concentrated on the expansion of the private rights accorded to information and knowledge owners (Boyle 1996; May 2000). Information or knowledge may have an existence outside the privately owned realm, but this is increasingly a residual category, recognised only when all conceivable private rights have been established. Thus, this line of criticism concentrates on the balance between private rights and public goods in the information society, suggesting there has been an over-compensation on behalf of private rights holders, to the detriment of the public realm. The enclosing dynamic disturbs previously legitimate settlements regarding private and public rights.

Any 'democratic' possibilities within the information society are overstated when seen from this perspective. For Julian Stallabrass, it

is not a matter of doubting the capabilities of the technology, which
has already been developed and is becoming cheaper all the time.
One should be deeply sceptical, however, about who will control
the information, how much it will cost, and to whom it will be sold.
Technological revolutions of the past parade their many broken
utopian promises.

(Stallabrass 1995: 10)

The democratic potential of information empowerment is merely that,
potential. In the real world of modern capitalism such developments are
marginal leaving the real dynamic of enclosure to continue unabated.
In the most extreme arguments from this standpoint, ICTs have rendered
people's control of their own environment and creativity the subject of
a totalised technology. George Spencer argues that human beings will
increasingly become irrelevant to the processes of production of goods
and delivery of services. The class of individuals unable to be part of
society because they lack any saleable skills will expand because this
revolution unlike earlier technological developments 'will not create a
demand for new forms of labour, for it will perform an increasing proportion of all activities itself' (Spencer 1996: 75). Given capitalism's logic,
no organisation will retain human labour once cheap and quick computing
can duplicate the task. The enhancement of control will render the system
more efficient, though immiserising large segments of the population.

In this extreme manifestation the enclosing position views the future
as akin to *Bladerunner*. Resistance may be possible but it is not part of the
systemic logic, it is abnormal. Or as Bettig suggests, though

> it is possible to find sites of resistance in cyberspace, the corporate
> forces bringing the information superhighway on line, following the
> logic of capital, undermine the liberatory potential in the technology.
> (Bettig 1997: 154)

Any potential for democracy is undermined and compromised by the
powerful and comprehensive enclosing dynamic. The information society
as intensified capitalism renders knowledge and information increasingly
as commodities and continually encloses the public realm through marketi-
sation and the search for new products and services. Those who regard the
potential of cyberspace as empowering, without positive and extensive
effort to make ICTs fit their hopes, are mistaken if the enclosing dynamic
defines the information society's developmental path. These fears fit well
with the characterisation of authoritarian technics in Mumford's work,

and represent the centralising and controlling tendencies of the mega-lopolis. But while these concerns are legitimate, they are as Mumford suggests only one side of technological development.

Democratic/disclosing

Arrayed against those who regard the information society as exhibiting a harsh enclosing dynamic are those who regard the information revolution as leading to expansive human empowerment. This competing perception suggests the information society exhibits a disclosing dynamic. Despite the opprobrium heaped on this position, it still garners considerable coverage (and indeed dominates discussion) outside the academy. For many Internet democrats, the information society will be a 'Jeffersonian demo-cracy', allowing the individual to prosper without undue interference from the 'authorities' (Barbrook and Cameron 1996). In some ways, with its regard for private property, a Jeffersonian information society might be better thought of as part of the enclosing dynamic. But 'democrats' like John Perry Barlow recognise that property in information resources is increasingly difficult to sustain and represents the type of control the infor-mation society is undermining. The disclosing dynamic is normalised and the enclosing tendencies represented as a threat or abnormality which either will or can be overcome: 'information wants to be free'. Information society therefore will be built upon interpersonal relations rather than through property relations (Barlow 1996). Self-acting, self-owning individuals will be equipped to enact social relations and engage with each other.

Information society, as democracy allows individuals to express themselves outside mass parties, outside class identities, it allows a new individuality. This is the result of the vast expansion in the information resources available for individuals to make such choices. Gone is the control of information by the expert, rather we can all access the infor-mation we need without the mediation of others. Furthermore,

> the good news is that information is leaky, that sharing is the natural mode of scientific discovery and technological innovation. The new information environment seems bound to undermine the knowledge monopolies which totalitarian governments convert into monopolies of power.
>
> (Cleveland 1985: 70–71)

Information will dissolve the threads of power. And Harlan Cleveland like others, argues that the flows of information will make hierarchies increasingly difficult to maintain even in formal democracies. Politics will be more concerned with people than geography, issues will become the mainstay of political interaction. In Manuel Castells' much cited trilogy on *The Information Age* he suggests that politics is coalescing around symbolic issues (the environment, human rights) which are driven by the disclosure of abuses previously obscured (Castells 1997: 309ff). In Mumford's terms a democratic technics, allowing 'communal self government, free communication between equals and unimpeded access to the common store of knowledge' is emerging through ICTs' ability to make such flows a reality where previously they might have been obstructed by the structures of industrial society.

In its most extreme manifestation, the disclosing perspective has a great deal in common with radical liberalism: restating conventional liberal views, Nicholas Negroponte asserts that the

> harmonising effect of being digital is already apparent as previously partitioned disciplines and enterprises find themselves collaborating not competing. A previously missing common language emerges, allowing people to understand across boundaries . . . but more than anything my optimism comes from the empowering nature of being digital.
>
> (Negroponte 1996: 230–231)

Conflicts will be resolved through knowledge, education and conversation. And again emphasising the Jeffersonian swing of the democratic/disclosing dynamic, Negroponte worries that for the Internet the 'only hazard is government in the form of politicians who want to control it' (Negroponte 1996: 234). Again, the conflicting dynamic (here of enclosure) is presented as abnormal and a danger which is at odds with the defining (and 'real') character of the information society. Mumford's human-scale technics, where the tools serve the individual's interest rather than the system's is implicit within such assertions. The information society in this view will allow knowledge-based power to flow down to communities and individuals rather than being centralised in government.

The continuing relevance of Lewis Mumford

Arguments regarding the character of the emerging information society can therefore be broadly divided into an enclosing and a disclosing position. While each recognises aspects of the other, these secondary tendencies are treated as abnormal and problematic. Thus, for the enclosing dynamic, society is continuing a dynamic of commodification, the rendering of everything as property, and though aspects of the information society seem currently to be allowing contrary developments, such instances are merely transitory or marginal. Here information society is characterised by enclosure and while this may be accepted or criticised, the systemic logic will defeat those abnormal practices which currently act to disclose information. Conversely, the disclosing position regards the dynamic towards enclosure as being merely a temporary hold-over from previous forms of social organisation. Enclosure will become outmoded and impossible allowing the real dynamic of the information society (disclosure) to triumph. Enclosure is seen as abnormal and often presented as a threat from old power bases, a threat that needs to be resisted but will be defeated in time because it runs contrary to the real dynamic of the information society.

Lewis Mumford's important insight is that actually these two dynamics are not contradictory: both are continuing elements of the history of technology and its social milieu. The move to control and enclosure (authoritarian technics in its systemic mode) exists alongside the tendency to disclosure (the possibility of democratic technics). For Mumford the history of technology has been a process of interaction and conflict between democratic and authoritarian technics. These technics have not been the result of specific technologies but are the product of the social, political and economic relations in which they appear, are developed and deployed. Thus, rather than one or other of information society's dynamics being abnormal, both dynamics must be regarded as parts of the character of the information society. The dominant ontology of normality and abnormality renders the nominated dynamic as the driving force behind the information society while the other is regarded as a problem, a resistance or a misapprehension of the 'logic of informationalism'. But while there may be dangers in authoritarian technics (the enclosing dynamic) these can be tempered through social and political practices. In essence Mumford's position regards the determinism and fatalism of the enclosing dynamic as simplistic (and possibly ideologically driven), while also recognising that the disclosing dynamic may underestimate the resistance (both explicit and systemic) to the possibilities it highlights.

Therefore, an account of the information society needs to accord to each dynamic an analytical importance which does not render the other as abnormal. This suggests that analysis of the information society has to encompass both dynamics: it needs to recognise the challenge of enclosure to disclosure and vice versa is not an abnormality but rather the way the global information society itself develops (as previous technological ages have developed). Indeed, to develop a political economy of the information society the implications of this dual dynamic need to be understood as a complex system: not as contradictory and problematic. To achieve this aim a reacquaintance with Lewis Mumford's analysis of the history of technology is not only useful, but also necessary if the sterile dispute between the enclosing and disclosing positions is to be left behind. It locates the information society in the long and eventful history of technology, and as importantly within a history of competing technics. Additionally it explains why currently at least neither the worst fears of the enclosing dynamic have been manifest, nor the best hopes of the 'democrats' have been achieved. By recognising this as part of an historic dialectic between authoritarian and democratic technics a complex and useful analysis of the new megalopolis, the invisible city of the information society can be developed.

Acknowledgements

I thank Richard Barbrook, Robin Brown Jonathan Nitzan and Jayne Rodgers for commenting on earlier versions of this chapter, but as always the remaining shortcomings are the author's own.

Notes

1   A useful account of Mumford's method (which however omits his debt to Hegel) can be found in Novak (1987).
2   In the original Feenberg is discussing both Mumford *and* William Morris as occupying a similar position with regard to the social construction of technology.
3   While I have mainly drawn this account from the article in *Technology and Culture* (Mumford 1964), the same argument is made in substantially similar terms in *Technics and Human Development* (Mumford 1966b: 234–242). Fores (1981) argues that in *Technics and Civilisation* (Mumford 1934) the term 'technics' is only a mask for technological determinism, but even if this is the case, in his later work this term is used to identify the social embeddedness of technology rather than its determination of history. Fores claims are forcefully put but take *Technics and Civilisation* as a single work and are not related to Mumford's post-war shift in ideas regarding the history of technology.
4   In *The Culture of Cities* (Mumford 1940) he briefly included an intermediary phase between megalopolis and necropolis, that of 'tyranopolis'. However, this was dropped when he returned to the subject in later works.
5   I owe this point to Robin Brown.

References

Barbrook, R. and Cameron, A. (1996) 'The Californian ideology', *Science as Culture* 26(1): 44–72.

Barlow, J.P. (1996) 'Selling wines without bottles' in P. Ludlow (ed.) *High Noon on the Electronic Frontier*, Cambridge, MA: MIT Press. Widely available on the Internet, at numerous sites.

Beniger, J.R. (1986) *The Control Revolution: Technological and Economic Origins of the Information Society*, Cambridge, MA: Harvard University Press.

Bettig, R.V. (1997) 'The enclosure of cyberspace', *Critical Studies in Mass Communication* 14(2): 138–157.

Boyle, J. (1996) *Shamans, Software and Spleens: Law and the Construction of the Information Society*, Cambridge, MA: Harvard University Press.

Casillo, R. (1992) 'Lewis Mumford and the organicist concept in social thought', *Journal of the History of Ideas* 53(1): 91–116.

Castells, M. (1997) *The Power of Identity* (The Information Age: Economy, Society and Culture: vol. 2), Oxford: Blackwell.

Cleveland, H. (1985) 'The twilight of hierarchy' in B.R. Guile (ed.) *Information Technologies and Social Transformation*, Washington, DC: National Academy Press.

Feenberg, A. (1991) *Critical Theory of Technology*, New York: Oxford University Press.

Fores, M. (1981) '*Technik*: or Mumford reconsidered', *History of Technology* 6: 121–137.

Goist, P.D. (1972) 'Seeing things whole: a consideration of Lewis Mumford', *Journal of the American Institute of Planners* 38 (November): 79–391.

Jamison, A (1995). 'The making of Lewis Mumford's technics and civilisation', *EASST Review* 14(1). Available: http://www.chem.uva.nl/easst951.htm#jamison (1 October 1998).

Kundnani, A. (1998) 'Where do you want to go today? The rise of information capital', *Race and Class* 40(2/3): 49–71.

Lasch, C. (1980) 'Lewis Mumford and the myth of the machine', *Salmagundi* 49 (summer): 4–28.

May, C. (1998) 'Capital, knowledge and ownership: the information society and intellectual property', Information, Communication and Society 1(3): 245–268.

May, C. (2000) *A Global Political Economy of Intellectual Property Rights: The New Enclosures?* (RIPE series), London: Routledge.

Miller, D.L. (1989) *Lewis Mumford: A Life*, Pittsburgh, PA: University of Pittsburgh.

Morley, J. (1989) 'International Symposium on Lewis Mumford – University of Pennsylvania, November 5–7, 1987', *Technology and Culture* 30(1): 122–127.

Mumford, L. (1924) *Sticks and Stones: A Study of American Architecture and Civilisation*, New York: Boni & Liveright.

Mumford, L. (1934) *Technics and Civilisation*, London: George Routledge and Sons.

Mumford, L. (1940) *The Culture of Cities*, London: Secker & Warburg.

Mumford, L. (1962) 'Apology to Henry Adams', *Virginia Quarterly Review* 38 (spring): 196–217.

Mumford, L. (1964) 'Authoritarian and democratic technics', *Technology and Culture* 5 (winter): 1–8. Reprinted in M. Kranzberg and W.H. Davenport (eds) (1972) *Technology and Culture: An Anthology*, New York: New American Library.

Mumford, L. (1966a) 'Technics and the nature of Man', *Technology and Culture* 7: 303–317.

Mumford, L. (1966b) *Technics and Human Development* (Myth of the Machine: vol. 1), New York: Harcourt Brace Jovanovich.

Mumford, L. (1968) [1962] 'Megalopolis as anti-city', *Architectural Record* (December). Reprinted in *The Urban Prospect*, London: Secker & Warburg.

Mumford, L. (1971) *The Pentagon of Power* (Myth of the Machine: vol. 2), London: Secker & Warburg.

Mumford, L. (1975) [1961] *The City in History: Its Origins, its Transformations and its Prospects*, Harmondsworth: Penguin.

Negroponte, N. (1996) *Being Digital*, London: Hodder & Stoughton.

Nitzan, J. (1998) 'Differential accumulation: towards a new political economy of capital', *Review of International Political Economy* 5(2): 169–216.

Novak, F.G. (1987) 'Lewis Mumford and the reclamation of human history', *Clio* 16(2): 159–181.

Robins, K. and Webster, F. (1999) *Times of the Technoculture*, London: Routledge.

Spencer, G (1996) 'Microcybernetics as the meta-technology of pure control' in Z. Sardar and J.R. Ravetz (eds) *Cyberfutures*, London: Pluto Press.

Stallabrass, J. (1995) 'Empowering technology: the exploration of cyberspace', *New Left Review* 211 (May–June): 3–32.

Ward, C. (1986) 'Introduction' in L. Mumford, *The Future of Techniques and Civilisation*, London: Freedom Press.

Winters, S.B. (1987) 'The achievement of Lewis Mumford – New Jersey Institute of Technology, Newark, October 14–17, 1985 – Conference Report', *Technology and Culture* 28(1): 106–112.

Chapter 7

Karl Polanyi

Kenneth S. Rogerson

Karl Polanyi (1886–1964) was an anomaly among economists. He considered himself an economic anthropologist. What may seem like an oxymoron in some circles seemed perfectly logical to him. Traditionally, economists in general have explained the world in theory (other things being equal) and thus in isolation from social factors. In fact, the advent of international political economy and those who followed it was seen as slightly revolutionary. Not many scholars do work in, for example, sociological economics, philosophical economics or cultural economics, though research in these areas has been extant for some time.

Though not a prolific writer, Polanyi was absolutely consistent. He spent his life's work describing and explaining how economic relations affect individuals and societal groups. His best known work (and, many would say, his masterpiece) was *The Great Transformation*, which examined the historical evolution of the industrial revolution and posited a rationale for many of the negative societal consequences that Polanyi believed plagued society because of it. Following a life theme, one of Polanyi's (1944: 249) major conclusions was: 'The true criticism of market society is not that it was based on economics – in a sense, every and any society must be based on it – but that its economy was based on self interest'. And, self-interest leads people to search for ways economic relations benefit only them, not taking into consideration if or how those relations can be detrimental to others.

As the body of Polanyi's work becomes more widely read, as well as commentary on it becomes more prevalent, it is worthwhile to revisit his theories and concepts in a new context. Polanyi's work focused on the industrial revolution. The end of the twentieth and the beginning of the twenty-first centuries have sometimes been referred to as the 'information revolution'.[1] Revolution means radical change. If both periods are indeed revolutionary, the difference between the two ages is simply one of the

mechanisms of change. In the late nineteenth century, mechanisation –
for Polanyi (1944: 33) a 'radical improvement in the tools of production'
– changed how people understood and used money, labour, markets and
people. In the late twentieth century, it has been argued that through even
more mechanisation (such as innovative communication and information
technologies) information flows have had a similar, globally relevant
effect.

One of the great dilemmas of the information age is the tension between
two dynamics: first, the tendency of information to be free flowing and not
to lose its value as it moves, and second, the tendency to want to control
that flow of information in order to profit from its value. In *The Great
Transformation*, Polanyi identified similar contradictions in the industrial
revolution, except the flowing material was capital instead of information.
In other words, a free market implies that capital can freely flow to the
place at which it makes the most profit. Any controls – by governmental
or commercial entities – are put in place to improve the chance to profit
from that flow. Polanyi spent a lifetime exposing what he felt to be the
negative consequences of this increasingly free-flowing capital. In the end,
he advocated a type of welfare state in which government intervention is
necessary to counteract these negative effects.

Given some of the issues of the information age – such as the digital
divide – as well as the arguments of the proponents and critics of these
issues, this chapter will analyse Polanyi's work and apply some of his
concepts to the information age. Of the many concepts Polanyi provides
in his analysis of economic relations, two stand out as carefully calculated
assumptions about the relationship between society and economy: first,
'embeddedness', how embedded economic relations are in society, and
second, 'the double movement', opposing forces which attempt to 'right
the wrongs' brought about by an unregulated market economy. Extending
this to the information age, I shall examine how information exchanges
and flows are embedded in society and, as a result of attempts to control
these flows, I shall analyse the landscape of existing and potential opposing
forces to this control.

There are three parts to this discussion. First, an examination of the
assumptions Polanyi makes about the industrial age and to what extent
those assumptions are applicable in an information age. Second, a
discussion about how the concept of embeddedness contributes to our
understanding of why this might be called the information society.
Finally, whether a double movement exists (I believe it does) and what its
potential consequences might be.

The Industrial Revolution and the Information Revolution

Polanyi approached his study of economic relations very historically. Writing in the late 1940s, he used examples from the mid-nineteenth century to support his contentions about the unequal beneficiaries of the industrial revolution. Polanyi believed that the bedrock of a market economy as understood at the turn of the century was what he referred to as *haute finance*, a global society of banks that worked with each other and governments. 'The motive of *haute finance* was gain; to attain it, it was necessary to keep in with the governments whose end was power and conquest' (1944: 11). This network has certain characteristics: privately owned enterprises, tied to, but mostly independent from, governments; a relatively balanced flow of goods and services based on the idea of supply and demand; an entrepreneurial spirit personified by extremely wealthy individuals who tried to exercise influence in various political and social situations; and little or no government regulation.

In addition, he maintains that the relationship between international finance and governments contributed to peace-making efforts. In order to maximise their profit, banks working with each other encouraged trade. 'Trade had become linked with peace. In the past the organization of trade had been military and warlike [pirates, explorers, slave traders, etc.]. Trade was now dependent upon an international monetary system which could not function in a general war. It demanded peace' (Polanyi 1944: 15).[2] Because of this connection between banks and the international system, there was tension between regulation of monetary interactions and military interactions. The only way to ease this tension was to replace a semi-regulated system with a freer market economy, which placed a barrier in the way of 'social progress' and 'social reform'. This is the crux of Polanyi's thesis. 'The origins of this cataclysm [the barrier] lay in the utopian endeavor of economic liberalism to set up a self-regulating market system' (1944: 29).

He continues that, following the implementation of the principles of a free market economy,

> an avalanche of social dislocation . . . came down upon England; that this catastrophe was the accompaniment of a vast movement of economic improvement; that an entirely new institutional mechanism was starting to act on Western society; that its dangers, which cut to the quick when they first appeared were never really overcome; and that the history of nineteenth century civilization

consisted largely in attempts to protect society against the ravages of such a mechanism.

(Polanyi 1944: 40)

As an antidote to the discomforts and social ills resulting from this system, Polanyi suggests state intervention (see Polanyi 1944; Myrdal 1960).[3] Government intervention is not designed to be intervention in the market, but intervention in society to offset the negative effects of the market. In other words, government must help those in society who are unable or unwilling to use market forces for personal gain and are thus negatively affected by them.

The information revolution has some striking parallels, but also some differences. Information has been socially and politically relevant for centuries. In a broadly historical overview, Harold Innis (1972) emphasised that the ability to control information and communications processes is vital to those seeking to attain and maintain political dominance.[4] Innis describes 'monopolies of knowledge' which 'developed and declined partly in relation to the medium of communication on which they were built' (1972: 166). For Innis, this implies that there is an abstract concept of 'knowledge' or information which, when used in the right way, can contribute to dominant status. But it has only been in the twentieth century – and even in the latter part of that century – that the gathering, storage, and analysis of information have fundamentally changed social and political interactions.

With the development of electronic forms of communication (telegraph, telephone, radio and television), the potential uses, the speed and the volume of information increased radically. New technologies (computers, computer networks such as the Internet, satellites, wireless communications) have provided even more opportunities for superlatives. These technologies were developed mostly in the western world and many of them in the service of war. As telegraph technology improved in Europe in the mid-1800s, international cooperation in the communications arena began with the formation of the International Telegraph Union (later International Telecommunication Union). Guglielmo Marconi was working on early prototypes of the radio. Various scientists, mostly from the United States and Europe, contributed to the development of the television. In the immediate post-Second World War era the US government focused on the development of new technologies. '[O]f particular importance was investment in new communications technology. Initially funded by the armed forces, the investment boosted the growth of the telecommunications industry and of television' (Holt 1995: 168).

Much of the innovation in information and communications technology at this time was motivated by the bipolar rivalry between the United States and the Soviet Union. Advanced methods of surveillance in the form of satellites, storage and retrieval of information, and an increasing ability to interpret that information was absolutely necessary to 'win' battles in the Cold War. 'The primary weapons of the Cold War were ideologies, alliances, advisors, foreign aid, national prestige – and above and behind them all, the juggernaut of high technology . . . Of all the technologies built to fight the Cold War, digital computers have become its most ubiquitous, and perhaps it most important, legacy' (Edwards 1996: ix). Edwards argues that the history of the computer is inextricably tied to 'the elaboration of American grand strategy' during the Cold War and that computers 'made much of that strategy possible' (1996: 2).

Though much of the technology was initially produced for specifically military uses, ideas such as the Advanced Research Projects Agency Network or ARPANET, which was created in 1969 by the US Department of Defense to facilitate the sharing of research information and findings, eventually became the backbone of what is known today as the Internet in the United States. Other types of communications technologies, like small satellite receivers, were initially intended for communications use within the military (see Harris 1994; Scott 1994).

As with the tension between banks and governments in the industrial revolution, there was a tension that pulled the development of information technology into the commercial arena. Especially in the post-Cold War period and with the expansion of the commercial aspects of the Internet in the mid-1990s, flows of information through new technologies came to take on some of the same characteristics that Polanyi saw in the industrial revolution. It is clear from the Internet startup boom, for example, that most people involved believed that the expansion of ways to use information was proof that the free market economy worked (see Ozawa *et al.* 2001).

But some hot-button issues of the information age – such as the digital divide, the commoditisation of information, and security and privacy – as well as the arguments of the proponents and critics of these issues, provide some insight into a more complex analysis of the revolution and its consequences.

In both the industrial and information eras, because of mechanisation, the speed with which people were able to accomplish tasks increased. From the cotton gin to assembly line, industrial age innovations were intended to expand the number of products or the amount of service available while decreasing the number of bodies and the amount of time needed. Computer processors, fibre optic cables, robotics and enhanced spectrum use are the spinning jennies of the information age, accomplishing many of the same things.

Second, the volume of products increased in the industrial age. Illustrating the point, Polanyi said (with his trademark negative undertone), that the basis of the industrial age was the development of the machine. 'Since elaborate machines are expensive, they do not pay unless large amounts of goods are produced. They can be worked without a loss only if the vent of the goods is reasonably assured and if production need not be interrupted for want of the primary goods necessary to feed the machines' (Polanyi 1944: 41). In the same way, the general belief is that the technological changes of the information age have resulted in an increase in the amount and types of available information. There is an 'endless amount of information online . . . Certainly, in terms of information sources, we . . . have [much more] to choose from than in the comparatively anemic world of radio and TV' (Shapiro 2000: 181–182).

Third, theoretically, the cost of both production and information decreased. Economic theory states that there is an equilibrium at which profit is maximised and that equilibrium comes when productions costs are at their optimal low while prices are at their optimal high, that is, the point at which consumers will believe it is still worthwhile to buy the product. The parallel in the information age is that, over time, the cost of generating, storing and analysing information has decreased. A simple look at the change in the cost of television, videocassette recorders or computers can support this assertion.

There are also some differences. The units used to describe each age – information and products – are not the same thing. They have different characteristics. For example, products may lose their value over time, but information tends to retain its value (see Braman 1989; Badenoch *et al.* 2000). Also, in the information age, the scope of change is exponentially beyond the scope of change of the industrial age. The ability to store such immense amounts of information could be akin to warehousing and distribution systems, but in much greater quantities. And the increased ability to analyse information does not seem to have a parallel, like better quality control or improved interactions between different types of businesses. These came later, some with the information revolution.

The next step, then, is to ascertain whether the concepts that helped Polanyi better understand the social, political and economic interactions of the industrial age – in this case embeddedness and double movement – provide a better understanding of societal relations in the information age.

Information embeddedness

Whether something is embedded in something else in a social sense is open to interpretation, depending on how the concept is operationalised. Polanyi believed that the economy was originally a 'point of intersection between lines of activities carried on by larger kinship groups' and did not substantially contribute to individual motives for action which 'spring as a rule from situations set by facts noneconomic – familial, political or religious' (1957: 71). In other words, for Polanyi, the concept hearkens to the time when people engaged in economic activity because they needed to in order to survive. This activity was also the source of a number of public goods since economic activity for the community led to the production of enough goods for societal subsistence. Polanyi believed that, over time, the economy became 'disembedded' from its social bases, that is, people engaged in economic exchanges simply for the benefit those exchanges bestowed upon the individual. This disembedding resulted in negative societal consequences because people began to engage in economic exchanges for personal gain at the expense of the larger societal need or potential advantage.

One of the important mechanisms for this change was the increased desire for commodities, which led, as one example, to the commoditisation of labour and land, things that were originally part of society and not bought and sold. Ultimately, this altered the nature of most transactions from purely social ones to principally economic ones. In fact, 'the Industrial Revolution . . . was utterly materialistic and believed that all human problems could be resolved given an unlimited number of material commodities' (Polanyi 1994: 40).

Polanyi termed this change the development of 'fictitious commodities'. He believed that labour, land and money had become a vital part of industry. 'But labor, land and money are obviously *not* commodities; the postulate that anything that is bought and sold must have been produced for sale is emphatically untrue in regard to them' (Polanyi 1994: 72, original emphasis). He continues by saying that labour is simply another way to describe common human activity which cannot be detached from the rest of life; land is simply another word for nature; and money is simply a token of purchasing power. None of them is produced for sale.

Can a similar conceptualisation be applied to the relationship between information and society? A discussion of the embeddedness of information in society may seem to be stating the obvious. Information is everywhere. But, Polanyi's radical notion that the economy was embedded in society may have been met with similar reactions in his day. And, in fact, just such a discussion about the embeddedness of information in society has been part of communications theory for some time.

Theoretically, Joseph Klapper summarised discussions about the effects of communication on social change, ultimately explaining the embeddedness of information in societal exchanges. To emphasise this point, he said that mass communication does not usually serve as a necessary and sufficient cause of something, but rather functions with other factors and influences. In other words the simple flow of information alone seems to have little effect in and of itself on the outcome of social interchanges. In Polanyi's words, it is embedded in other social relations, such as political, economic, cultural, or health-related discussions and actions (Klapper 1990: 12). Klapper believed that one common thread running through many theories of communication is that what people experience in their societal interactions is often mediated through channels of information – for example, the mass media, computers or the Internet. Information flows do have important consequences for individuals, for institutions and for society and culture, but these consequences must be understood as part of a greater context of social interactions.

How embedded information exchanges are in society is a key issue with one of the newest technologies: the Internet. Is it something new (that fundamentally changes the type of information we receive) or simply a means of disseminating more of what already exists? How much do people rely on the Internet as an exchange of information without really thinking about its impact on their lives?

Following the example of the commoditisation of labour, land and money above, some might argue that there has been a price on information for a long time.[5] Though not disputing this, others believe that new forms of technology are significantly expanding the tendency of information to be commoditised.

> The availability and use of specialized forms of information are becoming more and more concentrated. Because of high production costs in relation to prospective consumers, this kind of information commodity is expensive, generally sought by and beneficial to the world's wealthy and well educated.
>
> (Comor 1998: 218)

Is this also a fictitious commodity? If, indeed, information is so embedded in our societal relations, it is highly probable that Polanyi would place information in the same category as labour, land and money. Following Polanyi's logic, this does not mean that information should not be commoditised – indeed, he believes that 'labor, land and money markets *are* essential to a market economy' (Polanyi 1944: 73, original emphasis) – but commoditisation should not be the *only* way that the potential social (economic) value of information (or labour, land or money) is recognised.

Given the prevalence of free market ideology, the commoditisation of information is inevitable and once it takes place, Polanyi's logic leads us to believe that there will be negative social consequences. One example of this is what has come to be known as the 'digital divide', or the discrepancies on a global or national level between those who have access to information through new technologies and those who do not. The debate addresses both how to define the concept as well as the indicators of it, crossing the lines of academic, political, societal and policy-making circles.

Contending policy views suggest three ways of looking at the digital divide. The first is that the digital divide does not exist. One author explained this view, contending that the growing availability of inexpensive computers and free Internet service providers 'clearly call into question the presence of a significant digital divide in America . . . The marketplace is doing more than an adequate job of providing computing technologies to Americans' (Thierer 2000: 3). Mick Brady suggests that discussions about the digital divide are simply taking attention away from more pressing problems. 'The idea that people in third world countries should be encouraged to walk a half-day to have an Internet experience is obscene. The digital divide is not a crisis. World hunger, wars, AIDS, and environmental decay are crises. When the Internet can solve those problems, maybe everyone needs to have a computer' (Brady 2000).

The second view suggests that the digital divide exists, but that the situation is improving. In a preface to the October 2000 report of the Department of Commerce, former US Secretary of Commerce Norman Mineta wrote that the United States, for example, 'is moving toward full digital inclusion' as 'the number of Americans who are utilizing electronic tools in every aspect of their lives is rapidly increasing' (US Department of Commerce 2000: ii). Yet, Mineta confirmed that 'the digital divide still remains. Not everyone is moving at the same speed' and some 'groups are progressing more slowly' than others. Others have a similar optimistic view on an international level (see Mesquita 2001: 29–31).

The third view suggests that the digital divide exists, but rather than improving, the situation is worsening. Roger Crockett argued this view, stating that 'the notorious digital divide isn't closing' (Crockett 2000: 56) suggesting that the gaps between the haves and the have-nots are growing, rather than decreasing. At this point, the digital divide is a sensitive political issue because data can be found supporting all three views.

However, given his propensity to see inequalities, Polanyi would probably believe that it does exist, based on the inequalities that have existed since the industrial revolution. Though some like Brady (2000) would use this argument to claim the nonexistence of the digital divide, Polanyi's work leads to a different conclusion. The natural flows of information in society have been disembedded, similar to the case with natural economic interactions. In most instances, the disembeddedness has come from commoditisation, in others from political desires for power and control. The results of this are socially disruptive. In fact, 'modern society continues to protect itself against the forces that undermine its social solidarity and threaten to distort its relationship to the natural environment' (Baum 1996: 6). The inequalities evident in the digital divide are akin to the inequalities that Polanyi saw in the industrial revolution. The similarities between the time periods do not negate the existence of a digital divide. Rather they lend support to an understanding of the rise of a backlash against the inequities, the process of which Polanyi identified in *The Great Transformation*.

Double movement

This societal desire to protect itself is what Polanyi called the 'double movement', or 'the action of two organizing principles in society, each of them setting itself specific institutional aims, having the support of definite social forces and using its own distinctive methods' (1944: 132). For Polanyi, the two opposing forces present during the industrial revolution were first, the principle of economic liberalism, self-regulating markets supported by the owning and trading classes, and second, the principle of social protection, the safeguarding of society by social forces that seek to protect the people, their land and their culture.

There are two characteristics which tend to define how the double movement works: first, entrepreneurial dynamism, 'entrepreneurial decisions over the structure and location of output and technology have had extreme external impacts. Decisions such as those which have rendered obsolete fixed capital or labour skills, eradicated product lines of industries, or altered the development path of communities or regions have been made upon the basis of entrepreneurial calculation'. And second, interventionist drift, the tendency of social groups to orchestrate actions which are a necessary response to the utopian experiment of market capitalism (Stanfield 1986: 121).

Political interventionism resolves the tension between the two movements. The manifestation of this intervention is socialism, which, according to Polanyi, is 'the tendency inherent in an industrial civilization to transcend the self-regulating market by consciously subordinating it to a democratic society' (1944: 234).

There are examples of this double movement in the information age. One of the more prominent and global is the tension between the industrialised world and the less industrialised world (encouraged by the Soviet Union) in UNESCO in the 1970s and 1980s. At a conference in Montreal in 1969, UNESCO political bodies began to focus on Third World complaints about dependence on communications and information infrastructures centred in advanced industrial societies. The less developed countries charged that a huge imbalance existed in the global distribution and control over communications resources and that major western wire services monopolised world news coverage, creating a one-way flow of information (see Hachten 1987, especially Chapter 3).[6]

The most visible actions in this movement came with a call for a New World Information and Communication Order (NWICO). The UNESCO General Conference in 1970 adopted a resolution inviting member states to 'take the necessary steps, including legislative measures, to encourage the use of information media against propaganda on behalf of war, racialism and hatred among nations' (Giffard 1989: 20).

The idea was given impetus with the onset of microelectronics and the development of computerised satellite transmission. Studies on the one-way flow of messages and the monopoly of television programming (approximately 75 per cent of which were produced by the United States) made a strong impression and caused alarm among UNESCO member states. This led many of the non-aligned countries to demand a contextual analysis of the growing inequality and imbalance which widened the information gap between countries (Gonzales-Manet 1988: 34).

The NWICO became important for non-western countries because it provided an arena for criticisms levelled against western nations, the United States in particular. The subject matter was also less concrete than, for example, complaining of imperialistic tendencies in the military or economic realms. The demands of those desiring a NWICO were not new. In fact, they were synthesised from what many felt were decades of inequality in the global flow of information. Some demands included: the flow of news between developed and developing countries should be equal and balanced; western media should use more stories from foreign news sources; and an international body should be developed to ensure the flow of news between developed and developing nations (Frederick 1993:

169–170). Those nations with more democratic and capitalistic leanings supported the free flow of information while those adhering to communist, and some socialist, ideologies stood behind the premise of the importance of state control.

Thus, the 'battlecries' within UNESCO became very distinct for the different global regions. For the United States and its western allies, it was 'free flow of information and a free press'. For the Soviet Union and its allies, it was 'greater state control of information'. For the developing world, it was 'cultural imperialism', manifest in the calls for a new information order. One of the avenues designed to explore this tripartite tension of ideologies was the MacBride Commission (Harley 1993).[7] Discussions at commission meetings revealed both the perspectives of the United States as well as other countries' attitudes toward the United States and other western nations. It was clear that both the United States and the Soviets were so caught up in an ideological war that they ignored the position of the developing world. One delegate proposed that the commission should address the new information order. The Russian delegate immediately 'attacked the proposal on the basis that no such thing existed. "There is no such order . . . You can't just announce that there is a new world order and then put it into effect".' The US delegate agreed, saying 'he had just recently heard the term and he didn't know what it meant . . . It appeared to be nothing but an empty phrase, and if it was not, why was it so hard for Third World countries to explain it?' (Harley 1993: 52). The ultimate result was the withdrawal of the United States and Great Britain from UNESCO in 1983 and 1984.[8]

Yet, the commission did permit coalitions of countries which had not before been heard on the subject, a chance to voice their views. Ultimately, as Polanyi would have predicted, they organised themselves in opposition to the negative effects of the information revolution on their societies. The MacBride Commission Report

> showed the intersection of and radical changes in the industrial, educational, information, and cultural sectors. It uncovered contemporary processes which the transnational power elites have tried to conceal and misrepresent. It points out the growing intensification of class and ideological struggle confronting the world. It draws attention to the crisis and questioning of information monopolies, as well as the methods of control exercised by a few Western countries for more than a century. It is an implicit denunciation of the monopolistic modernization schemes and the new forms of capitalist domination.
> (Gonzales-Manet 1988: 33)

The protests that arose in the debate were never fully addressed by those who felt taken advantage of and they have arisen again in the late 1990s with the expansion and diffusion of the Internet. At the first conference on Internet and Society at Harvard University, discussions centred on the seeming American 'cultural imperialism' of the information age. 'The French have taken to the barricades, certain that an Anglo-centric World Wide Web is poised to bulldoze their culture into oblivion . . . In Europe, Asia, Africa and truly around the globe, concern is mounting over America's overwhelming influence in cyberspace' (Herschlag 1996). Though some things have changed over time – such as the increasing number of languages and cultures represented on the Internet – the western world, and the United States in particular, is still in a dominant position, in fact, it may even be a 'leadership' position (Hart 2000). In fact, most of the 'Top 100' websites visited are controlled by already existing corporations (Culnan 1999), and almost all are based in the United States.

There are those who believe that the diffusion of information and its accompanying technologies comes with a price, and these groups have organised themselves to oppose the negative consequences they see in the information age. From the NWICO in UNESCO to recent protests at the various meetings of the World Trade Organization to the introduction of the Linux operating system as an alternative to Microsoft's Windows operating system (Lessig 2001: 54–55), there are varying levels of social mobilisation.

In summary, both of the characteristics Polanyi used to describe the double movement – entrepreneurial dynamism and interventionist drift – as well as the political intervention that bridges the two, are evident in the history of the NWICO debate. Entrepreneurial dynamism is seen in the dialogue between East and West, to which a North–South conversation played a secondary role. Any decisions in the field of information or communications developments were based primarily on military or security rationale. High-level political elites (entrepreneurs) made decisions without much thought about the consequences for those outside of the purview of the situations or events. The interventionist drift was manifest in the organisation of the Group of 77 and the Non-Aligned Movement, which, though initially addressing economic inequalities, gained momentum and cohesion in UNESCO while discussing information issues. UNESCO itself was the institution that provided an opportunity for political intervention on behalf of those subject to the unintended consequences. Initiatives such as the International Program for the Development of Communications (IPDC) provided a way to address some of the concerns of the developing world. The purpose of the IPDC was to reorient UNESCO discussions and actions about communications away from ideological debates over new orders and toward the technical development of communications infrastructures in developing societies. The programme was supported by the United States and was thus accused of having ulterior motives, but there were some successes. For example, the IPDC was instrumental in the creation of the Pan African News Agency (PANA), which served as the continent's first interregional communications system (Coate 1988: 36). Polanyi's concepts do provide a coherent way to understand the dynamics and interrelationships in this example of the double movement in the information age.

Comparison and conclusion

One of the clearly possible conclusions from this discussion is that the information age is simply an extension of the industrial age. The flows of the information revolution are only a logical next step in the mechanisation of the industrial revolution. In fact, this reading is too simplistic. The information revolution is not a subset of the industrial revolution, but has made it even broader. There is a uniqueness in what is flowing. The flow of information is different than the flow of capital. Though they share some characteristics, capital flows are only one type of information flow.

There are three areas of comparison between the two ages that support the relevance of Polanyi to an analysis of an information age. First, the concepts that Polanyi used to understand the industrial revolution; second, his suggestion of intervention as a response to the negative consequences he saw; and third, the nature of information itself.

First, concepts such as the embeddedness of information in society, the commoditisation (fictitiously) of information, and the existence of a double movement in the information revolution, do provide an improved understanding of the nuances specific to societal relations in this age of increased information flows.

Information has always been embedded in societal interactions. It has only been disembedded to a certain extent. Since information markets do not have exactly the same characteristics as product markets, simply commoditising information does not change the fact that it can be used in myriad other ways. A person may pay for access to an after-hours stock tip and then simply pass it on via phone or email for free and the value is not lost. A used product may have some value, but the speed with which it can be passed on is not comparable and it is probable that its condition will not be as good as when it was first purchased.

In addition, the existence of a double movement is not as easy to see in the information age because of a simple paradox. People who would eschew the uses of technology for ideological reasons, may turn to it for other reasons. Though now removed, witness the Taliban Online, which was in English, no less (Chroust 2000). Also, groups who would organise protests against a number of issues are often likely to protest a specific issue and not the imbalance of information flows or the ownership of information dissemination methods. For example, a community may demand that child pornography be taken off the Internet, but not that the Internet itself be shut down.

Second, Polanyi advocates government interventionism as the solution. In fact, one of the great dilemmas of the information age is the tension between two dynamics: first, the tendency of information to be free-flowing and not to lose its value as it moves, and second, the tendency to want to control that flow of information in order to profit from its value. Economists claimed a similar tension in the industrial age. Markets want to be free so that concepts such as Adam Smith's 'invisible hand' will encourage the market to its optimal equilibrium (Polanyi 1944: 43–45). What has happened in both cases is that governments have intervened. But, the trajectory of that intervention has varied both with the type of government and the state of technological advancement within a country.

For example, on 28 December 1995, CompuServe, a global Internet access provider, blocked access to more than 200 sexually explicit computer discussion groups world-wide. This action came in response to an announcement by the German government that it would investigate the company on charges of violating the country's strict pornography laws. While CompuServe examined the legal ramifications of the investigation, it denied access to more than 4 million customers, 3.75 million of whom lived outside German borders. CompuServe, the German government and members of German society desired to exercise influence over information – CompuServe as a private organisation encouraging the free flow of information for its profit; the German government desiring to restrict a small portion of that flow on political and legal grounds; and segments of German society calling for restrictions for moral reasons (Rogerson and Thomas 1998: 427).

On a different level, governments can be proactive, implementing programmes that aim to bridge the digital divide, such as Singapore's attempt to become a paperless society and projects in other countries to connect all schools.

Finally, both economic interactions and flows of information have the propensity to be universal. 'The substantive connotation of economic is universal because all human groups must somehow secure their livelihood by turning nature's laws to their advantage' (Stanfield 1986: 50). Another way this is put in the information age is the 'borderless' nature of information. Polanyi understood that whether or not some interactions transcend the parameters put in place by the constraints of a market, there are times when some interactions *should*. If the nature of information is to flow, then the Polanyian response would be that it should flow for the public good and that governments *should* be actively involved in providing that space for it to flow. This involvement needs to be more than simple intervention in technical issues. It must be at the nexus of technological-societal-political-economic relations.

The long-term effects of the information revolution remain to be seen. The true Polanyi-esqe analysis must take place in a decade or two in order to have a similar sweeping, historically based view of the societal impact of information flows. At the same time, the conceptual tools Polanyi consciously provided, as well as some that we have been able to use in new ways, help us to gain insight into the dynamic relationships between the economy, polity, society and information flows.

Postscript: Karl Polanyi

To emphasise how he felt about his research, Polanyi spent his life as a social activist. Born in Vienna in 1886 and raised in Budapest under the mantle of the Austro-Hungarian empire, Polanyi considered himself a political activist but non-partisan, working for various societal groups designed to improve social conditions. He was 'openly idealist and voluntarist' and he wrote about what he saw as participation in politics simply for the sake of political participation. 'The last ten years have been devoted to an exploration of Hungary, and in doing this important work we have overlooked the fact that there is no mention of enthusiasm on even the most accurate economic and social map of Hungary, and without it the map is but a piece of paper' (Levitt 1990: 31). Polanyi did not 'play an active role in . . . politics', but he and his wife Ilona Duzynska 'were enthusiastic supporters of the socialist municipal administration of Vienna whose pioneering housing, cultural and community developments attracted international attention' (McRobbie and Levitt 2000: viii).

As life and politics changed in Austria he was forced to flee to London and eventually, just preceding the Second World War, to the United States. He penned *The Great Transformation* from 1941 to 1943. Because of his wife's communist activities, she could not get a US visa after the war and they lived in Toronto while he taught at Columbia University. He died in 1964.

His bibliography is short but rich. His masterpiece is *The Great Transformation: The Political and Economic Origins of our Time* (1944). He followed this with two very different examinations of how the concepts he laid out in *The Great Transformation* are applicable in differing contexts: *Dahomey and the Slave Trade* (1966) and *The Livelihood of Man* (1977, compiled and edited by Harry Pearson after Polanyi's death). He was also principal editor and author in a volume of essays entitled *Trade and Market in the Early Empires: Economies in History and Theory*. Finally, he wrote two articles worth mentioning. First, 'Ports of trade in early societies' (1963) in the *Journal of Economic History*. Second, a review of Shakespeare's *Hamlet* (1954) for *The Yale Review*, which provides one of the more intimate looks at his life and his work.

Notes

1 The literature on revolutions will not be invoked here. Suffice it to say that the simple question of whether this age is truly revolutionary or not will be left for a future generation. Shapiro (2000) believes it is. May (2002) believes it is not.
2 Polanyi (1944: 15) acknowledges that the traditional approach to understanding the expansion of capitalism has been 'anything but peaceful'.
3 There are varying degrees of state intervention from simple regulatory regimes to a complete welfare state.
4 Harold Innis applied the idea of the importance of communication specifically to the construction of empires and, thus, implies that the control of communication is extremely relevant in the political arena.
5 Think, for example, of ancient spies and other such intelligence-gathering operations that were subject to corruption and disloyalty for economic reasons.
6 With the creation of the UN Conference on Trade and Development (UNCTAD) in 1964, those nations who felt that the decision-making process in the United Nations was unfair to lesser developed countries proposed a New International Economic Order (NIEO). The NIEO, through the Non-Aligned Movement and the Group of 77, became a call for greater economic equality throughout the world.
7 The commission was named for Seán MacBride, an Irish journalist and politician respected for his diplomatic skills.
8 Even though the United States is still not a voting member, it maintains a three-person observer mission in Paris and continues to fund programmes on a voluntary basis. Great Britain rejoined in 1997.

References

Badenoch, D. *et al*. (2000) 'The value of information' in M. Feeney and M. Grieves (eds) *The Value and Impact of Information*, London: Bowker Saur.

Baum, G. (1996) *Karl Polanyi on Ethics and Economics*, Montreal: McGill-Queen's University Press.

Brady, M. (2000) 'The digital divide myth', *E-Commerce Times* 4 August. Available: http://www.ecommercetimes.com/news/viewpoint2000/view-000804-1.shtml (5 January 2002).

Braman, S. (1989) 'Defining information: an approach for policymakers', *Telecommunications Policy* (September): 233–242.

Chroust, P. (2000) 'Neo-Nazis and Taliban Online' in *28th International Political Science Association World Congress*, 1–5 August 2000, Quebec, Canada.

Coate, R.A. (1988) *Unilateralism, Ideology, and US Foreign Policy: The United States In and Out of UNESCO*, Boulder, CO: Lynne Rienner.

Comor, E. (1998) 'Governance and the "commoditization" of information', *Global Governance* 4(2): 217–233.

Crockett, R.O. (2000) 'How to bridge America's digital divide', *Business Week* 8 (May): 56.

Culnan, M. (ed.) (1999) 'Privacy and the top 100 web sites: report to the Federal Trade Commission', created June 1999. Available: http://www.msb.edu/faculty/culnanm/gippshome.html (1 March 2002).

Edwards, P.N. (1996) *The Closed World: Computers and the Politics of Discourse in Cold War America*, Cambridge, MA: MIT Press.

Frederick, H.H. (1993) *Global Communication and International Relations*, Belmont, CA: Wadsworth.

Giffard, C.A. (1989) *UNESCO and the Media*, New York: Longman.

Gonzales-Manet, E. (1988) *The Hidden War of Information* (trans. Laurien Alexandre), Norwood, NJ: Ablex.

Hachten, W.A. (1987) *The World News Prism: Changing Media, Clashing Ideologies*, Ames, IA: Iowa State University Press.

Harley, W.G. (1993) *Creative Compromise – The MacBride Commission: A Firsthand Report and Reflection on the Workings of UNESCO's International Commission for the Study of Communication Problems*, Lanham, MD: University Press of America.

Harris, J.F. (1994) 'In electronic battlefield training exercise, fog of war', *Washington Post* 24 April: A18.

Hart, M. (2000) *The American Internet Advantage: Global Themes and Implications of the Modern World*, Lanham, MD: University Press of America.

Herschlag, Miriam (1996) 'Cultural imperialism on the new; policymakers from around the world express concern over U.S. role', press release, Harvard University Conference on the Internet and Society, 27 May.

Holt, R. (1995) *The Reluctant Superpower: A History of America's Global Economic Reach*, New York: Kodansha International.

Innis, H.A. (1972) *Empire and Communications,* Toronto: University of Toronto Press.

Klapper, J.T. (1990) 'The effectiveness of mass communication' in D.T. Graber (ed.) *Media Power in Politics*, 2nd edn. Washington, DC: Congressional Quarterly Press.

Lessig, L. (2001) *The Future of Ideas: The Fate of the Commons in a Connected World*, New York: Random House.

Levitt, K. Polanyi (ed.) (1990) *The Life and Work of Karl Polanyi*, Montreal: Black Rose.

McRobbie, K. and Levitt, K. Polanyi (eds) (2000) *Karl Polanyi in Vienna*, Montreal: Black Rose.

May, C. (2002) *The Information Society: A Sceptical View*, Malden, MA: Blackwell.

Mesquita, R.L. (2001) 'The digital age after the first wave', *New Perspectives Quarterly* 18(1): 29–31.

Myrdal, G. (1960) *Beyond the Welfare State: Economic Planning and its International Implications*, New Haven, CT: Yale University Press.

Ozawa, T. , Castello, S. and Phillips, R.J. (2001) 'The Internet revolution, the "McLuhan" stage of catch-up, and institutional reforms in Asia', *Journal of Economic Issues* 35(2): 289–299.

Polanyi, K. (1944) *The Great Transformation*, Boston, MA: Beacon Press.

Polanyi, K. (1957) 'Aristotle discovers the economy' in K. Polanyi, C.M. Aresnberg and H.W. Pearson (eds) *Trade and Market in Early Empires*, Glencoe, IL: Free Press.

Rogerson, K.S. and Thomas, G.D. (1998) 'Internet regulation process model: the effect of societies, communities and governments', *Political Communication* 15(4): 427–444.

Scott, W.B. (1994) 'Satellites key to "Infostructure"', *Aviation Week and Technology* 14 March: 57–58.

Shapiro, A.L. (2000) *The Control Revolution: How the Internet is Putting Individuals in Charge and Changing the World We Know*, New York: Century Foundation.

Stanfield, J.R. (1986) *The Economic Thought of Karl Polanyi: Lives and Livelihood*, New York: St Martin's Press.

Thierer, A.D. (2000) 'How free computers are filling the digital divide', *The Heritage Foundation Backgrounder no. 1361*, 20 April. Available at: http://www.heritage.org/library/backgrounder/bg1361es.html (1 November 2001).

US Department of Commerce (2000) 'Falling through the Net: toward digital inclusion', report by the US Department of Commerce, October. Available: http://www.ntia.doc.gov/ntiahome/fttn00/contents00.html, (15 February 2002)

Chapter 8

Elmer Eric Schattschneider

Robin Brown

Compared with the volume of writing on the economics, sociology and culture of the information society, relatively little attention is given to its political implications. Authors tend to address the politics *of* the information society rather than how informationalisation changes political life. This chapter argues that one way of understanding how politics is changed in a more transparent society can be found in the work of E.E. Schattschneider (1892–1971). Particulary in *The Semisovereign People* (1960) he explores the dynamics of politics in a shrinking world where publicity becomes a key weapon. While Schattschneider's concern was with the implications of the nationalisation of American political life during the twentieth century, his key concepts of scope, visibility and the socialisation of conflict are of even greater value in an era of satellite television, the Internet and the mobile phone. Schattschneider makes a trenchant case for the autonomy and importance of political activity both analytically and morally, defending the claims of democracy and unmasking the limitations of contemporary politics. In doing this he provides an important counter-balance to the tendency to reduce the politics of the information society to sociology or economics.

The chapter falls into three main sections. The first part places Schattschneider and his work in the context of American political science's early-twentieth-century debate on the nature of democracy and politics in the age of mass media. The second introduces the main ideas of *The Semisovereign People*, the contagiousness of conflict, scope, visibility and privatisation. The third section looks at these ideas in the context of emerging trends in global politics.

Schattschneider and Democracy

In his introduction to *The Semisovereign People*, David Adamany observes that Elmer Eric Schattschneider 'was first a democratic philosopher and partisan, and he was a political scientist only so far as being so was useful in discovering and advocating means by which the people could get control of the government' (Adamany 1975: xxvii). This helps to explain some of the paradoxes about this scholar. A highly respected figure, he was elected President of the American Political Science Association in 1957–1958 and chaired their working group on the US party system. Yet he spent most of his career not as a research professor but primarily as a teacher of undergraduate students at Wesleyan University in Middletown, Connecticut, where he devoted much of his time to political activity and urging his students to become involved in political life. He wrote hugely influential studies of the workings of the US political system but increasingly abjured academic justifications and explanations in favour of an almost aphoristic turn of phrase. One footnote in *Two Hundred Million Americans in Search of a Government* reads in its entirety: '[p]olitical science: a mountain of data surrounding a vacuum' (Schattschneider 1969: 8).

The explanation for these paradoxes can be found in the transformation of American political science in the first four decades of the twentieth century. From the beginnings of the republic the study of politics in the United States was marked by a commitment to the education of the citizen in the values of the constitution. But by the 1920s in parallel was a growing scepticism about the ability of the citizen to participate constructively in political affairs and political scientists were coming to a vision of the field more concerned with scientific investigation than teaching (Ball 1995). This scepticism was summarised in Walter Lippmann's *Public Opinion* (1922). Lippmann argues that 'the omnicompetent citizen' of democratic myth who can formulate an informed judgement on the issues of the day can no longer be found in a complex urban society. In such a society politics is conducted via the mass media rather than via personal experience or direct debate but the modern newspaper is too fickle and uninformative to perform this key democratic role. The outcome must be the recognition that modern society must increasingly depend on experts rather than citizens (Lippmann 1922: 173, 229). John Dewey's *The Public and its Problems* (1927) was intended as a rebuttal to Lippmann but came to much the same conclusions. Political science increasingly emphasised the role of interest groups and the unimportance of the mass of citizens in shaping political decisions. This pluralism initially represented an analytical approach but came to be seen as a normative account of the political system. While writers accepted that the US system did not work in the way suggested by the constitution it was argued that it still provided a viable model of democratic practice (Almond 1990: 82–83).

While Schattschneider accepted the need for a more empirically based analysis of the political system he rejected the elitism and detachment of the new political science. Like Dewey, Schattschneider distinguished between democracy as an idea and as a system of government. As a system of government democracy was about the consent and judgement of the citizens. His idea of democracy was more radical: it 'begins as *an act of imagination about people*. For this reason democracy is a doctrine of social criticism' (Schattschneider 1969: 46, original emphasis). This idealist vision of politics was in tension with his realist view of the world. Although in contemporary political science realism is assumed to be a theory of international politics that focuses on the state, in the 1930s United States it had a much broader application. Realism stood for a view of politics that was motivated by the clash of interests between groups – in the context of the Great Depression the groups that mattered were rich and poor (e.g. Niebuhr 1932). While Schattschneider leaned towards a pluralist view in analytical terms, he was suspicious of the normative version of the idea. In a much quoted phrase the 'flaw in the pluralist heaven is that the heavenly choir sings with a strong upper-class accent' (Schattschneider 1960: 34–35). Politics was a constant struggle between groups and anyone who set out to make democracy more thoroughgoing and effective needed to understand this.

There is little here that would suggest that Schattschneider can shed much light on the dynamics of the information society. But Schattschneider drew on a strand of thinking about American politics that goes back to the beginning of the republic; this emphasises the importance of space as a factor that shapes the dynamics of political life. Montesquieu and Rousseau argued that responsible government was possible only in small territories. As a polity grew larger its government could be made to work only by involving smaller and smaller groups (Montesquieu 1748; Rousseau 1762). In their defence of the constitution the Federalists argue that the extent of the United States protects against the influence of minorities and faction-alism and by doing so makes republican government feasible (Madison *et al.* 1987: 122–128). However, given the size of the United States there must be sufficient intercourse between its parts to ensure some coherence. In the words of James Madison, defending the power of the federal govern-ment to establish post offices and post roads, '[n]othing which tends to facilitate the intercourse between the States can be deemed unworthy of the public care' (Madison *et al.* 1987: 279). This connection between space, communications and politics recurs throughout the history of the United States to the extent that Carl Hovland observed that 'in the United States communication is a substitute for tradition' (cited in Carey 1999: 87). The problem that motivated Lippmann and Dewey was the mismatch between the scale of economic and social life and the fact that democratic political practices had their origins in local communities. How could democracy be made to function on the same scale? As Dewey put it, '[p]olitical and legal forms have only piecemeal and haltingly, with great lag, accommo-dated themselves to the industrial transformation' (Dewey 1927: 111, 114). Here we can see a distinct parallel between the problems of American politics and the contemporary challenges of globalisation. Where Dewey treated this as 'primarily an intellectual problem', Schattschneider's realism gave a different perspective but it was understanding this relation-ship between social change and political that motivated Schattschneider (Dewey 1927: 126).

Apart from *The Semisovereign People* Schattschneider published three major books. The first, based on his doctoral thesis, *Politics, Pressure and the Tariff* (1935), examined the way in which pressure groups had been able to shape the Smoot-Hawley Tariff Act of 1930. Relying on an analysis of the 20,000 pages of testimony taken before Congress, he was able to see how government was effectively handed over to private interests in a way that accelerated the globalisation of the Great Depression. *Party Government* (1942) explained why US political parties were simultane-ously central to the working of the system and unable to effectively manage the pressures of interest groups. His final book *Two Hundred Million Americans in Search of a Government* (1969) is a collection of essays that in part foreshadows the concerns of later generations of poli-tical scientists – the interdependence of domestic and international politics and the relationship between political change and political language – but it is in the *The Semisovereign People* that we find Schattschneider's most penetrating attempt to make sense of the changes around him.

Politics as a fight

The subject of *The Semisovereign People* was the way in which American politics was adapting to an increasingly national pattern. Parties and interest groups were having to adapt their strategies to take account of this new reality and the result was the growing importance of national politics and the decline of a politics centred on cities, states and regions. If this is the focus, Schattschneider's approach to his subject matter appears to be rather oblique. The book opens with the following passage:

> On a hot afternoon in August, 1943, in the Harlem section of New York City, a Negro soldier and a white policeman got into a fight in the lobby of a hotel. News of the fight spread rapidly throughout the area. In a few minutes angry crowds gathered in front of the hotel, at the police station, and at the hospital to which the injured policeman was taken. Before order could be restored, about four hundred people were injured and millions of dollars worth of property were destroyed.
>
> (Schattschneider 1960: 1)

The point of this is that

> all conflict has some elements that go into the making of a riot. Nothing attracts a crowd so quickly as a fight. Nothing is so contagious. Parliamentary debates, jury trials, town meetings, political campaigns, strikes, hearings, all have about them some of the exciting qualities of a fight; all produce dramatic spectacles that are almost irresistibly fascinating to people. At the root of all politics is the universal language of conflict.
>
> (Schattschneider 1960: 1–2)

This is simply a statement of political realism, that politics is fundamentally a matter of conflict rather than of reason or justice, but it is Schattschneider's next step that is so productive.

> Every fight consists of two parts: (1) the few individuals who are actively engaged at the centre and (2) the audience that is irresistibly attracted to the scene. The spectators are as much a part of the over-all situation as are the overt combatants. The spectators are an integral part of the situation, for, as likely as not, the *audience* determines the outcome of the fight. The crowd is loaded with portentousness because it is apt to be a hundred times as large as the fighting minority, and the relations of the audience and the combatants are highly unstable.
>
> (Schattschneider 1960: 2)

It then follows that the

> first proposition is that the outcome of every conflict is determined by the extent to which the audience becomes involved in it. That is, the

outcome of all conflict is determined by the scope of its contagion
... every increase or the reduction in the number of participants,
affects the result.

(Schattschneider 1960: 2)

This means that the 'most important strategy in politics is concerned with
the scope of conflict'. The reason for this is that

it is extremely unlikely that both sides will be reinforced equally as
the scope of the conflict is doubled or quadrupled or multiplied by a
hundred or a thousand . . . It follows that conflicts are frequently won
or lost by the success that the contestants have in getting the audience
involved in the fight or in excluding it.

(Schattschneider 1960: 3–4)

How and why groups came to be involved in conflict was at the heart
of Schattschneider's work. Although very conscious of the power of
economic interests in society he argued in *Party Government* that there was
a 'law of imperfect political mobilization', that it was organisation that
converted potential interests into real political factors. Assuming that one
could simply read off politics from economic interests was a mistake
(Schattschneider 1942: 19, 21, 33). Parties and interest groups operated
strategically to mobilise support in order to wage their chosen conflicts
by attempting to involve supporters or exclude opponents. Those already
involved attempted to control who joined in, but as conflicts became more
visible such control became more difficult.

Although conflict was contagious political groups sought to control
it by shaping which conflicts became dominant. They attempted to define
the lines of cleavage that politics would be waged along.[1] Choosing to
fight on one issue rather than another would influence who would choose
to become involved. Lines of cleavage come to be reflected in the organisational and institutional forms of political life. The organisations and
rules that structure political life tend to make it easier to pursue some issues
rather than others. Some 'issues are organized into politics while others
are organized out'. The existence of organisations dedicated to particular
goals will tend to elevate the visibility of those conflicts: 'organization is
the mobilization of bias' (Schattschneider 1960: 69). Classically political
parties were organised on class lines. This tended to organise class conflict
in but organise out other lines of cleavage such as gender conflict.

This process was not only organisational but also ideological because
of ideas held in the community about the proper sphere of politics. Changes

in these ideas contributed to the changing balance between what he termed the privatisation and socialisation of conflict. In Schattschneider's view there were an infinite number of conflicts in society but relatively few of them were incorporated into the political system. To socialise a conflict was to redefine it from being a private matter to one that was the responsibility of the broader community – and in particular the state. Whereas any conflict can draw in more participants, socialisation implies a change in the status of a conflict so that the involvement of other agents comes to be seen as legitimate.

> Universal ideas in the culture, ideas concerning equality, consistency, equal protection of the laws, justice, liberty, freedom of movement, freedom of speech and association tend to socialize conflict. These conflicts tend to make conflict contagious; they invite outside intervention in conflict and form the basis of appeals to public authority for the redress of private grievances.
>
> (Schattschneider 1960: 7)

In contrast, ideas such as 'free enterprise' tended to keep conflict privatised and to exclude external intervention.

In a longer term perspective there was a tendency for the boundary of the public sphere to expand. Schattschneider discusses both the rise of the US labour movement and the civil rights movement in terms of a movement of conflicts from a hidden private realm to the public. As he comments, government 'in a democracy is a great engine for expanding the scale of conflict' (Schattschneider 1960: 12). The growth of government has enormously strengthened the tendency to the socialisation of conflict. This has worked in parallel with processes of industrialisation, urbanisation and nationalisation that have changed society and 'all but destroyed the meaning of the word "local"' so that the 'visibility of conflict has been affected by the annihilation of space which has brought into view a new world' (Schattschneider 1960: 13).

Schattschneider identified this logic at work in the development of labour relations in twentieth-century USA and in the civil rights movement. What had been defined as private matters became redefined as public as the scope of the conflict grew. In one sense any expansion of scope implies a degree of socialisation – more people are involved – for instance employers and trade unions become involved in collective bargaining. In a stronger sense socialisation implies the intervention of public authorities. Thus in the history of twentieth-century labour relations we can see a trajectory by which trade unions initially sought more government intervention in order to level the playing field. By the end of the century they were calling for 'free collective bargaining' – attempting to reprivatise conflict in order to protect a bargaining position that was being threatened by new forms of regulation of labour relations. Even at the beginning of the twenty-first century, despite two decades of neo-liberalism, these relationships have been reprivatised only to a limited extent. It is also worth noting here that we can see the efforts of feminism in similar terms converting issues of gender relations from a private matter to a public one (Elshtain 1993).

Thus, the central concern of *The Semisovereign People* is the dynamics of conflict. Two distinct but related logics are at work. First, the tension between the expansion of the scope of conflict and its limitation. Second, the way in which some conflicts are privatised or socialised. As we shall see in the next section, the information society, by facilitating the diffusion of information on a global basis, creates new possibilities for political strategy through the globalisation of conflict. The issue of the privatisation of conflict takes on a greater significance in an era of neo-liberalism than it did when Schattschneider originally wrote the book.

Schattschneider's work helped to shape the study of parties, interest groups and social movements but, although *The Semisovereign People* continues to be cited, its arguments have been incorporated into the conventional wisdom of the field (Mair 1997). This incorporation has tended to blur some of the distinctive elements of Schattschneider's approach. Two of these elements deserve mention. The first is the importance of the spatial metaphor at the heart of the *The Semisovereign People*. During the 1980s and 1990s there was a renewal of interest in spatial approaches to social science, for instance in the work of Henri Lefebvre, Edward Soja and David Harvey (Lefebvre 1991; Soja 1989; Harvey 1989). Schattschneider's work supplies a theory of politics that sits easily with such theorising and adds an additional dimension to it. Later work in political science has tended to lose sight of this although it has been recognised in political geography (e.g. Adams 1996). The spatial dimension can be taken literally in the way in which the information society permits people who are physically remote to become involved in conflict. It can also be taken to deal with the linguistic and organisational features of society that tend to exclude particular individuals and groups from decisions.

The second neglected element is the sophistication of Schattschneider's implicit social theory. Part of the reason why Schattschneider's writing is so accessible (and often entertaining) is that he dispenses with many of the justifications and explanations expected in academic writing; he has little interest in complex philosophical justifications. Despite this underpinning *The Semisovereign People* is a view of the political world that is distinctly structurationist in tone: agents pursue strategies that lead to outcomes that structure the context that they act in. The distinctive lines of conflict that shape the political landscape are not 'natural' but reproduced by existing patterns of organisation and strategy hence potentially open to challenge. In his willingness to combine both material and discursive conceptions of politics Schattschneider's ontology is reminiscent of Elias, Bourdieu or Giddens. Schattschneider eschewed such discussion but his work is free of the structural-functionalism that marked much of the political science of the period. His concern with politics as a practical activity allowed him to maintain a focus on the agent and its strategy. His picture of the world both allows for the continuities of political practice and makes clear how change can occur.

Schattschneider in the information age

The remainder of this chapter takes up ways in which Schattschneider's ideas can cast light on the way in which information technologies affect the politics of scope. The issues raised by the changing scope of US political life in the twentieth century in some ways foreshadow the politics of globalisation in a mediated world. Although these comments focus on the relationship between globalisation and the politics of scale it should be emphasised that this logic applies just as much to political life on the micro-scale.

In discussing the growing importance of national politics in US political life Schattschneider emphasises the expansion of the power of the federal government in producing this effect. What he does not notice is that this process took place in parallel with the growth of national broadcasting media, first radio, then television that tended to increase the visibility of national politics. Schattschneider's account of the transformation of American politics through its nationalisation raises another intriguing point of relevance in an era of globalisation: the transformation of US political life took place with marginal constitutional changes. Changes at the level of strategy had fundamental significance for the overall topography of American politics.

In a world where the means to gather and disseminate information are becoming more pervasive than ever before, the issue of political scope is pushed to the fore. As the visibility of conflicts increases, more people, in more places, have the potential to become involved. Historically, each innovation in communications has changed the patterns of politics.

As technologies have increased the visibility of conflict – from the pamphlets of the Reformation to the mass circulation press in the nineteenth century to radio and television – political elites and institutions have been forced to adapt to the new strategic possibilities. This is not to imply that technology *determines* the pattern of politics but given Schattschneider's understanding of politics as a fundamentally conflictual behaviour political actors can be expected to try and take advantage of technological innovations. Schattschneider's theory would suggest that groups that are disadvantaged at the current scope of politics would initially embrace the new technology. Dominant actors would seek to limit its impact but if this was not feasible they would have to adapt their own strategies to the new environment.

The 1980s and 1990s saw the introduction of a whole slate of new technologies that change the ways in which information is gathered, communicated and processed. The Internet attracts the most attention but this is only one of a number of new developments. The declining size and cost of video cameras provides pictures from trouble spots, the increasing number of telephones, both fixed and mobile, makes the dissemination of information easier. The emergence of multi-channel television disseminated via satellite and cable provides a growing variety of audio-visual products relative to the few nationally controlled channels provided by terrestrial analogue television. Behind these technologies lies a pervasive fall in the costs of communication as a result of the boom in the construction of satellite and fibre optic telecommunications networks (Finel and Lord 2000). Even in Afghanistan, one of the very few areas on the planet where numbers of telephones have fallen since 1990, we can be treated to a constant commentary on events via satellite phone equipped journalists (International Telecommunications Union (ITU) 2001; Stern 2001).

These technological developments have taken place at the same time as a shift in the political context. Democratisation and liberalisation, the decline in the relative prominence of security issues for many countries and the growth of international governance structures have produced a more open global political field. Many governments are more relaxed about sharing information and see the benefits of working with international and non-governmental organisations (Taft 2000). These political developments have changed political agendas and produced new forms of conflict.

The result of the new communications environment and the new agenda is that politics is pervaded by struggles over scope and the socialisation of conflict. We are living in a period when the scope of political life is constantly contested.

Easier access to information and the mobilisational possibilities that result make it easier to expand the scope of conflicts. This creates new possibilities for strategic action and political organisation. At its simplest this means protesters in a non-English-speaking country carry English placards for the benefit of the international media to mobilise international support. Emails mobilising support for a movement fall into the same category. The Zapatistas converted their struggle from one that concerns only a few people in one of the most marginalised areas of Mexico to one that attracts international attention and support. This provides an example of how technology facilitates the expansion of scope and by doing so changes the political outcomes by placing a constraint on the Mexican government (Knudson 1998).

The emergent anti-globalisation movement indicates struggle over the governance of global issues. Conflicts over global governance include efforts to defend national autonomy against the encroachment of external influences whether economic, cultural or explicitly political. An alternative line of cleavage is between those who seek to bind national actors to global standards of economic regulation or human rights. However, politics in a globalising world is not simply about 'global issues'; there are changes in the way in which domestic politics is practised. It becomes easier to reach out across space to challenge the scope imposed by elites or the convention that politics is defined and limited by national borders. This logic is made clear in the way in which human rights groups make use of what has been called the 'boomerang' strategy: seeking support from foreign governments to put pressure on their own government (Keck and Sikkink 1998: 36).

If some people want to expand the scope of conflict then others want to control it. This can be done in two ways. The first is to ensure that people are unaware of potential issues or actual conflicts. Making sure that politics is something that happens out of the public view limits who can be involved. The second strategy is to construct a political discourse that legitimises the involvement of some people but not others. In international terms the classic expression of this is the distinction between citizens, who have a voice, and foreigners, who are excluded. Both of these strategies have been robbed of some of their effectiveness by recent developments. The diffusion of technologies tends to undermine attempts to deny information. The almost universal adoption of economic policies that promote access to information technologies and to information as necessary for success merely reinforces these trends.[2] Discourses of 'them' and 'us' lose some of their force in an era when politics is influenced by the universal liberal conception of norms. The discourses of sovereignty and identity are balanced by the discourse of prosperity, global governance and human rights that tend to undermine discourses of identity based on notions of 'them' and 'us' (Finnemore and Sikkink 1998).

The possibilities of expansion of scope that are at work in an era of cheap and pervasive communications are enormous but Schattschneider's argument also raises the question of the socialisation of conflict. Socialisation implies not simply the occasional use of strategies that expand scope but also the institutionalisation of that expanded scope. In *The Semisovereign People* the limit of socialisation of conflict was through the involvement of the federal government. In a globalising world the possibilities for socialisation expand. Even in 1960 Schattschneider could write:

> One of the most remarkable developments in recent American politics is the extent to which the federal, state, and local governments have become involved in doing the same kinds of things in large areas of public policy, so that it is possible for contestants to move freely from one level of government to another in an attempt to find the level at which they might try most advantageously to get what they want. This development has opened up vast new areas from the politics of scope.

Conflicts can be socialised at the level of the state or at the level of global governance. Changing the scope may produce a conflict over different forms of socialisation. Are conflicts best managed at local, national or global levels? To return to the earlier example of labour relations, should these be managed at a national or a global level? The emergence of global and regional governance mechanisms not only provides a choice of socialisations but also calls into question national differences in the ways in which conflicts are managed.

The whole discourse of human rights represents a socialisation of conflict on an enormous scale. How private individuals treat others and how states treat individuals have been globally socialised through the creation of international agreements. What the idea of human rights does is to remove certain kinds of behaviour, for instance slavery or torture, from the private sphere and socialise them on a global scale. The global nature of this governance regime imposes obligations on state governments. Normatively, if not always practically, this opens the way for anyone who has had their rights violated to appeal to the global community as a whole or part of it (Risse 2000). A second area of socialisation is the development of environmental policy. Over the past half-century the environment has become recognised as a political issue; it has become socialised at a national level. The emergence of global environmental regimes has created a new form of socialisation that can threaten existing national arrangements. The US decision to pull out of the 1997 Kyoto Protocol on greenhouse gas emissions can be seen in terms of a struggle between those who favour a national socialisation of the environmental issue versus those who prefer the outcomes that will emerge from a global socialisation. Regardless of the scientific content of the argument, expanding the scope to the global level gives a voice to powerful supporters of reductions in emissions of greenhouse gases (Anonymous 2001). The economic realm is a site of contestation because of the partial nature of the socialisation. Two trends operate simultaneously: the development of mechanisms of global governance and the movement towards neo-liberal models of economic policy. The former would inevitably create tensions between national models of regulation and a global regime. The latter, to some extent, pushes towards a reprivatisation of conflict limiting the sphere in which government plays an active and direct role. These issues are underlined by the emergence of anti-globalisation organisations. While the new governance mechanisms, such as the World Trade Organization (WTO) are manifestations of public authority, they appear to reprivatise some conflicts – for instance in the way in which North American Free Trade Agreement (NAFTA) disputes procedures appear to allow private interests to override public authorities (Naim 2000; Public Citizen n.d.). On the other hand these organisations create a new opportunity for concerned groups to influence outcomes. For instance in the case of opposition to the construction of dams campaigners have been able to exert influence on governments through targeting the World Bank (Khagram 2000).

Domesticating global politics?

As pressure groups mobilize to take advantage of new political opportunities there is a further intriguing parallel with developments in early-twentieth-century USA. Writing about the development of public relations as a distinct profession Stuart Ewen traces it to

> that moment when aristocratic paradigms of deference could no longer hold up in the face of modern, democratic, public ideals that were boiling up among the 'lower strata' of society. At that juncture, strategies of social rule began to change.
>
> (Ewen 1996: 13)

As the scope of politics changed, as political life came to be played out in the pages of the mass circulation press, new groups became relevant. As scope increased, strategy changed. Rule had to be exercised via the media and via strategies of publicity rather than secrecy. The realm of international politics was much less affected by these changes. Despite Woodrow Wilson's promotion of a more open diplomacy, it remained controlled by relatively small secretive groups (Eban 1998: 75–76). At the beginning of the twenty-first century the new international agenda along with the new communications environment of the Internet, mobile phones and satellite television is accelerating the opening up of international relations. This forces an increasingly multifarious range of agents: states, non-governmental organisations (NGOs), international organisations, companies, to come to terms with this changing context. As individuals and groups learn more about events, they can mobilise to change things. As the targets of this new activism come under pressure, they are having to change their strategies to cope with the new demands placed upon them.

Schattschneider's first book provides some suggestive insights into the vagaries of information and politics. In *Politics, Pressure and the Tariff* (1935) Schattschneider points out that one of the factors that shaped the Smoot-Hawley Act was that only some of the potentially interested parties were actually aware that hearings were being held and were able to monitor and attempt to influence the outcome (Schattschneider 1935: 164–166). The hearings were not secret or closed it was just that they weren't very well publicised. In *Party Government* (1942) he commented that

> A well-organized and well-staffed pressure group is able to get much information not easily available to people generally . . . on the basis of information so collected, an organized special interest is able to rally its supporters and to apply pressure when and where it can be employed most advantageously. A good intelligence service is therefore the basis of effective pressure tactics.
>
> (Schattschneider 1942: 201)

It might be suggested that in contemporary politics with its profusion of information-gathering and disseminating technologies, most activities take place in a quasi-public space. *Public* because they are not necessarily secret but *quasi*-public because of the likelihood that few people will notice that they are happening in time to do anything about them. The fact that information is available does not necessarily make it politically relevant. If the role of Schattschneider's pressure groups was to monitor events and then to mobilise their supporters, this is precisely the type of strategy that makes sense in the current information environment. This environment can increase the effectiveness of strategies based on publicity: an NGO can effectively republicise quasi-public events such as the negotiations on the Multilateral Agreement on Investment to a new audience and by doing so mobilise them to act (Deibert 2000). Rather than assuming that the impact of technology makes organisation irrelevant, this suggests that the fragmentation of media spheres will actually make organisation more important as a way of triggering political action.

Although it is NGOs and transnational advocacy networks who are normally identified as the beneficiaries of this new environment, we should be cautious about drawing conclusions from the record as it stands. In frequently cited cases of NGO impact on the international policy process, such as the Multilateral Agreement on Investment and the Seattle WTO Summit, the NGOs' success had much to do with the failure of their targets to attempt to tell their side of the story. The evidence is that governments, companies and international organisations are recognising the importance of publicity as an instrument of politics. In recent years a growing volume of policy-oriented writing has appeared arguing that communication is growing in importance as a tool of diplomacy. A growing chorus of voices is calling for greater prominence to be given to 'public diplomacy' (e.g. Leonard and Alakeson 2000; Arquilla and Ronfeldt 1999; Centre for Strategic and International Studies (CSIS) 1998). In an environment where the scope of conflict constantly threatens to expand one needs to be aware of groups that could join in and try and persuade them that their concerns are being addressed. The comment of Philip Gould on the modernisation of the British Labour Party comes to mind: 'in a modern media environment, competence and good communications are insepa-rable: you cannot have one without the other' (Gould 1998: 334).

Conclusion

Much writing about the information society slips into a structural account of processes of change that will produce either a society where democratic participation is the norm or one where surveillance eliminates the possibilities of action. For Schattschneider this is too crude a reading of politics. The political possibilities that exist in society are not a simple function of its social characteristics but of the political structures and strategies that exist. The political consequences of social change depend on how individuals and groups organise to deal with them. Reduced to the bare essentials his position is that politics matters: it is effective organisation, mobilisation and strategies that convert possibilities into realities. In this respect Schattschneider offers an account of the public realm that is both more pessimistic and more optimistic than the influential Habermasian version (Habermas 1989). It is more pessimistic since Schattschneider sees conflict as a permanent part of the human condition: more optmistic through the importance he attaches to political action. Accounts of the information society that ignore politics (and hence agency) misunderstand the nature of contemporary social change. Borrowing from Schattschneider allows us to understand both the possibilities and the limits of the new environment. *The Semisovereign People* provides explanations for political stability as much as change.

More substantively, in a globalising world the spaces of political life are constantly contested and *The Semisovereign People* provides a conceptual tool-kit for dealing with just these issues. We see how changes in scope change the nature of political strategies and political outcomes. To borrow a concept from social movement theory the ramifying consequence of the communications revolution is to change political opportunity structures. Social movement theory builds on Schattschneider's observation that 'people are apt not to fight if they are sure to lose'. A smaller world is one where individuals and groups have a greater opportunity to challenge the scope of political life and by doing so change its outcomes (Schattschneider 1960: 4; Tarrow 1998). This is not happening in a vacuum but on a terrain shaped by the outcome of previous struggles. The most strongly entrenched bias in the contemporary world are the lines of cleavage marked by the boundaries of states.

The Schattschneiderian perspective on the politics of the information age balances a liberal optimism with realism. Politics is fundamentally a matter of power and interests but in the information age everything becomes, potentially, public. This creates a tension with Kant's maxim that one should act in a way that accorded with all one's actions being public (Kant 1963 [1795]). Thus, in a world of publicity one can be constantly challenged about one's actions, their consistency with norms and declared objectives. As Debora Spar has shown, in a world of branding reputation becomes a vulnerability as much as an asset; companies become open to protest through fear of damage to the brand (Spar 1998). Yet Schattschneider's realism would not lead us to expect an unproblematic victory of liberal values. A world of publicity elevates the spin doctor to the top of the political pantheon. The age of information marks a change in the nature of power, not its abolition. But where the chosen weapon is information and imagery, the battlefield is more level than it would be if the weapons were more material.

Notes

1   Events in the wake of the terrorist attacks on New York and Washington on
    11 September 2001 provide a perfect example of this. Defining the conflict as
    being between freedom and terrorism or Islam and the west would influence
    what the stakes of the conflict were and who would be involved. Hence the
    importance of imposing one definition or the other.
2   For instance there is a growing literature on the implications of the information
    revolution for the Chinese authorities' ability to control information (see Duan
    2001; Hachigian 2001).

References

Adamany, D. (1975) 'Introduction' in E.E. Schattschneider, *The Semisovereign
    People: A Realist's View of Democracy in America*, new edition, New York:
    Holt, Rinehart & Winston.
Adams, P.C. (1996) 'Protest and the scale politics of telecommunications',
    Political Geography, 15: 419–441.
Almond, G. (1990) *A Discipline Divided: Schools and Sects in Political Science*,
    Thousand Oaks, CA: Sage.
Anonymous (2001) 'Global warming: oh no Kyoto', *The Economist* 7 April.
Arquilla, J. and Ronfeldt, D. (1999) *The Emergence of Noopolitik: Towards an
    American Information Strategy*, Santa Monica, CA: Rand Corporation.
Ball, T. (1995) 'An ambivalent alliance: political science and American democ-
    racy' in J. Farr, J.S. Dryzek and S.T. Leonard (eds) *Political Science in History:
    Research Programmes and Political Traditions*, Cambridge: Cambridge
    University Press.
Carey, J.W. (1999) 'Innis "in" Chicago: hope as the sire of discovery' in C.R.
    Ackland and W.J. Buxton (eds) *Harold Innis in the New Century: Reflections
    and Refractions*, Montreal: McGill-Queen's University Press.
Centre for Strategic and International Studies (CSIS) (1998) *Reinventing
    Diplomacy for the Information Age*, Washington, DC: CSIS.
Deibert, R. (2000) 'International plug "n" play? Citizen activism, the Internet and
    global public policy', *International Studies Perspectives* 1: 255–272.
Dewey, J. (1954) [1927] *The Public and its Problems*, Athens, OH: Swallow.
Donovan, R. and Scherer, R. (1992) *Unsilent Revolution: Television News and
    American Public Life*, Cambridge: Cambridge University Press.
Duan, Q. (2001) 'The information revolution, digital divide and regime transition
    in China', paper presented at the International Studies Association Convention,
    Chicago.
Eban, A. (1998) *Diplomacy for the Next Century*, New Haven, CT: Yale University
    Press.
Elshtain, J. (1993) *Public Man and Private Woman*, Princeton, NJ: Princeton
    University Press.
Ewen, S. (1996) *PR! A Social History of Spin*, New York: Basic Books.

Finel, B. and Lord, K. (eds) (2000) *Power and Conflict in the Age of Transparency*, New York: Palgrave.

Finnemore, M. and Sikkink, K. (1998) 'International norm dynamics and political change', *International Organization* 52: 887–917.

Gould, P. (1998) *The Unfinished Revolution: How the Modernizers Saved the Labour Party*, London: Little, Brown.

Habermas, J. (1989) *The Structural Transformation of the Public Sphere* (trans. T. Burger with F. Lawrence), Cambridge: Polity Press.

Hachigian, N. (2001) 'China's cyber-strategy', *Foreign Affairs* 80: 118–133.

Harvey, D. (1989) *The Condition of Postmodernity*, Oxford: Blackwell.

International Telecommunications Union (ITU) (2001) *Focus: Teledensity*. Online. Available: www.itu.int/sg3focus/teledensityA.htm (9 September 2001).

Kant, I. (1963) [1795], 'Perpetual peace' in L. Beck (ed.) *Kant on History*, Indianapolis, IN: Bobbs-Merrill.

Keck, M. and Sikkink, K. (1998) *Activists Beyond Borders: Advocacy Networks in Global Politics*, Ithaca, NY: Cornell University Press.

Khagram, S. (2000) 'Toward democratic governance for sustainable development: transnational civil society organizing around big dams' in A. Florini (ed.) *The Third Force: The Rise of Transnational Civil Society*, Washington, DC: Carnegie Endowment for International Peace.

Knudson, J. (1998) 'Rebellion in Chiapas: insurrection by Internet and public relations', *Media, Culture and Society* 20: 507–518.

Lefebvre, H. (1991) *The Production of Space*, Oxford: Blackwell.

Leonard, M. and Alakeson, V. (2000) *Going Public: Diplomacy for the Information Society*, London: Foreign Policy Centre.

Lippmann, W. (1997) [1922] *Public Opinion*, New York: Free Press.

Madison, J., Hamilton, A. and Jay, J. (1987) [1788] *The Federalist Papers*, Harmondsworth: Penguin.

Mair, P. (1997) 'E.E. Schattschneider's *The Semisovereign People*', *Political Studies*, 45(5): 947–954.

Montesquieu, J. (1989) [1748] *The Spirit of the Laws*, Cambridge: Cambridge University Press.

Naim, M. (2000) 'Lori's war', *Foreign Policy* 118: 28–55.

Niebuhr, R. (1932) *Moral Man and Immoral Society*, New York: Charles Scribner.

Public Citizen (n.d.) 'Another broken NAFTA promise: challenge by U.S. corporation leads Canada to repeal public health law'. Online. Available: http://www.tradewatch.org/nafta/cases/ethyl.htm (28 May 2001).

Risse, T. (2000) 'The power of norms versus the norms of power: transnational civil society and human rights' in A. Florini (ed.) *The Third Force: The Rise of Transnational Civil Society*, Washington DC: Carnegie Endowment for International Peace.

Rousseau, J. (1968) [1762] *The Social Contract*, Harmondsworth: Penguin.

Schattschneider, E.E. (1935) *Politics, Pressure and the Tariff*, New York: Prentice Hall.

Schattschneider, E.E. (1942) *Party Government*, New York: Holt, Rinehart & Winston.

Schattschneider, E.E. (1960) *The Semisovereign People: A Realist's View of Democracy in America*, New York: Holt, Rinehart & Winston.

Schattschneider, E.E. (1969) *Two Hundred Million Americans in Search of a Government*, New York: Holt, Rinehart & Winston.

Soja, E. (1989) *Postmodern Geographies: The Reassertion of Space in Critical Theory*, London: Verso.

Spar, D. (1998) 'The spotlight and the bottom line', *Foreign Affairs* 77: 7–13.

Stern, C. (2001) 'War boosts popularity of satellite telephones', *Washington Post* 20 November: E1.

Taft, J. (2000) 'Non-governmental organizations: the voice of the people', *US Foreign Policy Agenda*, March. Available: http://usinfo.state.gov/journals/itps/0300/ijpe/pj51taft.htm (17 April 2000).

Tarrow, S. (1998) *Power in Movement: Social Movements and Contentious Politics*, Cambridge: Cambridge University Press.

Chapter 9

Raymond Williams

Des Freedman

Raymond Williams (1921–1988) was one of Britain's outstanding social and cultural analysts. The son of a railway signalman, he won a scholarship to study at Cambridge University and went on to teach working-class students in adult education. His books on *Culture and Society* (1958) and *The Long Revolution* (1961) opened up an anti-elitist approach to culture that emphasised the expressive contributions made by those traditionally written out of cultural history: the poor and the exploited. Together with his contemporaries, Richard Hoggart and E.P. Thompson, his work was a key part of the development of the academic disciplines of cultural and media studies in the UK in the 1960s. Turning his back on one of the established canons of British intellectual life, he challenged the notion that culture was an elite pastime referring solely to the fine arts and insisted instead that culture was 'ordinary', that it emerged out of the soil of every-day life. The study of culture, therefore, required anthropological as much as aesthetic skills and sensitivity to the history, traditions and daily practices of working people.

As professor of drama at Cambridge in the 1970s, Williams' intellectual range was outstanding and his writings on politics, literature, philosophy, drama, television and technology earned him a reputation as one of the leading radical critics of his day. His engagement with Marxism at the time led to his articulation of 'cultural materialism', a modification of what Williams saw as the economism of the concept of a determining 'mode of production'. For Williams, culture, media and language were as *productive* as the institutions and processes typically attributed to the economic 'base' of society and as vital in securing the production and reproduction of everyday life.

This chapter focuses on Williams' analysis of the development of communication technologies and his critique of technological determinism contained in three pieces of writing: his short book on *Television: Technology and Cultural Form* (Williams 1974), his historical account of the evolution of media, 'Communication technologies and social institutions' (Williams 1981) and the chapter on 'culture and technology' in *Towards 2000* (Williams 1985). While only a tiny part of Williams' overall work, the dangers that he identified concerning determinism and technophilia have been accentuated by the ongoing infatuation with the transformative power of the Internet and the theorising of an 'information society'. Williams constantly stressed the indeterminacy and contingent nature of technological development, unlike the shrill prophets of today who are happy to ascribe specific consequences to the growth of computer networks. For example, Nicholas Negroponte of MIT's Media Lab confidently predicts the end of national sovereignty in an age of global flows of 'bits' as 'landlords will be far less important than webmasters. We'll be drawing our lines in cyberspace, not in the sand' (Negroponte 1998). George Gilder (1995) is convinced that the 'centrifugal' force of the Internet will necessarily lead to the collapse of 'all monopolies, hierarchies, pyramids and power grids of established industrial society'. Even a more balanced commentator like Anthony Giddens, theorist of the 'Third Way', argues that the 'communications revolution has produced more active, reflexive citizenries than existed before' (Giddens 1999: 73).

Williams, writing before the popularisation of the Internet, confronts such *determined* positions. He is passionate about the possibilities of technological innovation but insists that the development, take-up and use of technologies are all shaped by the social relations of the world into which they enter. There is no 'natural' course of development, no predictable shape that a technology will assume but instead a conflict between the capacities of particular innovations and the priorities of the most powerful groups. That means that there is a future to struggle over and Williams provides us with the intellectual armour both to challenge current profit-led decisions about technology and to press for an alternative, democratic vision of communications.

Technological determinism

Williams describes technological determinism as an

> immensely powerful and now largely orthodox view of the nature of social change. New technologies are discovered, by an essentially internal process of research and development, which then sets the conditions for social change and progress. Progress, in particular, is the history of these inventions, which 'created the modern world'.
>
> (Williams 1974: 13)

Technological development, therefore, is seen to be an autonomous process whereby the inner logic of a particular technology unravels in a predictable, often inevitable, fashion and changes the world into which it is born. This is the idea that the discovery of the printing press necessarily led to the Enlightenment, that telegraphy led to the industrial revolution and that the Internet has led to an information age. While rejecting this false causality, Williams recognises the simplicity and hegemonic power of these propositions.

> The basic assumption of technological determinism is that a new technology – a printing press or a communications satellite – 'emerges' from technical study and experiment. It then changes the society or sector into which it has 'emerged'. 'We' adapt to it because it is the new modern way.
>
> (Williams 1985: 129)

What is lacking from such accounts is any notion of social power, interaction or intention. For determinists, an efficient and sophisticated technology will ultimately impose its own discipline and its own patterns over and above the efforts of specific agents to use technology for partic-ular purposes. Williams, on the other hand, seeks to restore social context to the process of innovation and to assess the extent to which technologies are called into being through the needs and desires of corporations, states, groups or individuals. The question then becomes one of who has developed the technology, in whose interests, for what purposes, for which audiences, and with what consequences?

In pursuing these issues, Williams came into direct conflict with the influential arguments of Marshall McLuhan that mass media provide a sensory extension of the human body. McLuhan believed that the global, interactive and instantaneous possibilities of new communications technologies, in particular, had helped to resurrect the organic nature of speech-based communications. The sheer psychic power of technologies like satellites and television was such that

> Our conventional response to all media, namely that it is how they are used that counts, is the numb stance of the technological idiot. For the 'content' of a medium is like the juicy piece of meat carried by the burglar to distract the watchdog of the mind.
>
> (McLuhan 1998 [1964]: 18)

Williams acknowledged the appeal of McLuhan's discussion of specific media forms but accused him of a formalism that erased all social and historical context from discussions of technology. In McLuhan's writing, '[a]ll media operations are in fact desocialised; they are simply physical events in an abstracted sensorium' (Williams 1974: 127). Furthermore, if there is an inner sensory logic to communication technologies as McLuhan claimed, then attempts to change the uses and to modify the effects of these technologies will be doomed, a fatalistic position with which Williams entirely disagreed.

Williams was certainly wrong to suggest that McLuhan's ideas would have only a limited shelf-life – after all, it is the latter's books that are being reprinted and restored to academic reading lists to 'make sense' of contemporary digital developments. But Williams also pointed out that McLuhan's thesis, as an example of seeing technology as the driver of history, would be constantly renewed, making it all the more important to grasp technology as 'at once an intention and an effect of a particular social order' (Williams 1974: 128). It is to this objective that we now turn.

Four statements on technological development

> [A] technology is always, in a full sense, social
>
> (Williams 1981: 227)

Williams distinguishes between *technique* and *technical invention* as the application and development of particular skills (in laboratories, workshops or Silicon Valley basements) and the social institution of the *technology*. He describes the latter as 'first, the body of knowledge appropriate to the development of such skills and applications and, second, a body of knowledge and conditions for the practical use and application of a range of devices' (Williams 1981: 227). Williams is particularly interested in the process by which a technical invention becomes an 'available technology' (Williams 1985: 130), in other words the decisions about which inventions to develop, invest in and manufacture (if we may still use the word). Far from a technique unravelling along its own internal logic, it is the behaviour of real individuals in particular historical circumstances that shapes the transformation of an innovation into a technology.

Williams illustrates this point by discussing how advanced tribal societies developed writing systems that reflected their increasing complexity as trade expanded and tasks were specialised in order to meet these changing circumstances. Progressive development of the technology both depended on and reinforced the further expansion of trade (Williams 1981:

228). Similarly, the rise of the popular press in the nineteenth century depended on innovations in printing and paper production that Williams argues were 'specifically sought' by proprietors at the same time as it was 'closely bound up with the more general changes which were producing the conditions in which the new social and cultural form was necessary' (1981: 231). In other words, innovations were demanded by capitalist entrepreneurs, but these demands connected to the far wider social trans-formations implicated in the industrial revolution that created both the need and the space that the newspaper might satisfy and occupy. In each case, the development of a particular technology was bound up with profound social changes that, in turn, would be affected by the perfor-mance of that technology. For Williams, technology is a *relationship*: it is 'necessarily in complex and variable connection with other social relations and institutions' (1981: 227).

For example, Williams argues that broadcasting was not invented in a single flash of inspiration but developed during an extended process of technical experiment and innovation. What was crucial, however, and what organised these experiments into an available technology, was the desire for a medium that would complement the contradictory experience of new forms of urban life, based simultaneously on increased mobility and social atomisation. Williams describes radio and television as forms of 'mobile privatization' and he argues that broadcasting 'in its applied form was a social product of this distinctive tendency' (Williams 1974: 26). It offered the possibility of extending people's horizons, of stimulating their curiosity, of providing them with news from 'outside', but it did so by focusing on the family home as the centre of this communicative process. Williams links this to shifts in the social organisation of capitalism from the 1920s onwards: increased centralisation of production and decision making and therefore a loss of control over one's daily life results in growing investment in the private domain. Broadcasting proved to be a suitable technology to link the private with the public and, in so doing, helped to change the definition of both.

Williams' model of technological development appears to suggest that innovation is contingent on periods of social change – he writes that new systems of communication like photography, cinema and broadcasting were 'incentives and responses within a phase of general social transfor-mation' (Williams 1974: 18). Does this mean that significant changes in communication are necessarily linked to wider upheavals in society? Certainly, for Williams, the key forms of the newspaper were developed during times of crisis, for example in Britain during the Civil War, the industrial revolution and the two world wars when popular 'anxiety and controversy' called for new forms of ideological transmission and therefore new social institutions (1974: 21–22). Periods of technological stability may be interrupted by fierce challenges to the existing institutional arrangements that, while they may seem to have been internally generated through the rise of new technologies, are more importantly linked to wider social forces. Williams gives the example of the paperback book, an innovation that shook up the publishing industry, as a change partly induced through developments in printing technology but also as the result of 'determinate economic institutions [that] brought market considerations to a much earlier stage in the planning and writing of books'. As he puts it, 'the technology never does stand alone' (Williams 1981: 232).

> The moment of any new technology is a moment of choice.
> (Williams 1985: 146)

By this, Williams means that there is no predetermined form or function to communication technologies and that instead the eventual outcome of the process of innovation is related to the selections and preferences of human actors, not mechanical (or digital) systems. Consider the case of radio that was the subject of competing metaphors in its early development, either as a 'phone booth of the air' or as a 'newspaper of the air'. The eventual outcome of *broadcasting* was due less to technological factors than by the lobbying of the main US telephone company to keep the 'common carrier' network to itself (see Sawhney 1996). For Williams, this demonstrated that the decisions over which system to use 'were made on already existing political and economic dispositions in the societies concerned, since the technology, obviously, was compatible with any or all of them' (Williams 1985: 131).

The key issue here is that it is the choices of the most powerful groups in society that determine the shape of technologies, a situation that for Williams explains the gap between the potential and actual social benefits of communications technologies as they are increasingly subject to commercial considerations. Williams described this form of commodification as a 'counter-revolution, in which, under the cover of talk about choice and competition, a few para-national corporations, with their attendant states and agencies, could reach further into our lives' (Williams 1974: 151). So, for example, the democratic role of the press has been profoundly undermined by its structural reliance on advertising revenue that rendered, in several cases in the UK, working-class readerships of over 1 million to be economically unviable as they were the 'wrong' sort of readers, unattractive to advertisers. According to Williams, a technology 'which had promised both extension and diversity had, in these circumstances [a free market-oriented society], produced a remarkable and specific kind of extension (what came to be called the "mass public") and, by comparison with its own earlier stages, an actually reduced diversity' (Williams 1981: 232).

The development of television provides another example of how decisions on institutional structures that prioritise state or corporate interests limit the democratic capacity of technologies. First, Williams (1974: 25) argues that broadcasting systems were primarily devised as means of transmission and reception with little concern for the content that would be broadcast. The model of centralised transmission and privatised reception was adopted before there was a consensus on what kind of material should be shown. This relates to a further problem that, while television could transmit live events relatively cheaply, original content was far more expensive. According to Williams, it would have been logical to set up a 'socially financed system of production and distribution' (1974: 30) to offset the costs and ensure an adequate supply of funding for more expensive programmes. Instead, television has come to depend on advertising, sponsorship and insecure licence fees, a solution that Williams argues has led to under-investment in production and a lowering of cultural expectations.

Williams predicted that this dilemma would be repeated with the new information and communication technologies he saw being developed in the 1980s. While he envisaged alternative, socialist uses of cable and satellite systems to extend diversity and citizen participation, he warned that the introduction of new technologies on the lines of 'selective profit-taking' (Williams 1985: 148) would dissipate their democratic potential.

> Within existing social and economic conditions, the new systems will be installed as forms of distribution without any real thought of corresponding forms of production. New cable or cable-and-satellite television will rely heavily on old entertainment stocks and a few cheap services. New information systems will be dominated by financial institutions, mail-order marketers, travel agencies and general advertisers. These kind of content, predictable from the lines of force of the *economic* system, will be seen as the whole or necessary content of advanced electronic entertainment and information. More seriously, they will come to define such entertainment and information, and to form practical and self-fulfilling expectations.
>
> (Williams 1985: 146–147)

Narrow profit-led decisions about technological development shut down the range of possible uses for new technologies but also provide no guarantee as to the success of investment decisions. The current excess capacity in broadband systems and limited take-up in broadband services is a dramatic illustration of Williams' point that technological potentiality is not always matched by actual investment decisions and institutional forms.

> The sense of some new technology as inevitable or unstoppable is a product of the overt and covert marketing of the relevant interests.
>
> (Williams 1985: 133)

That technological determinism is seen as 'common sense' is the result of a strategy pursued by dominant groups in order to secure an acceptance of their institutional models for particular innovations. Precisely because competing models are technically possible, corporations are forced to attempt to convince investors, regulators and the public that the opposite is true: that there are no alternative paths and that resistance is futile because technological development is predetermined. Technological determinism, therefore, is a discursive means of highlighting novelty and paving the way for structural changes that are then seen to be necessary. For example, cable and satellite technologies, 'because they can be *represented* as socially new and therefore as creating a new political situation, are in their commonly foreseen forms essentially paranational' (Williams 1985: 139, emphasis added). This provides a technical, rather than a political, justification for loosening existing national regulatory mechanisms in ways that will benefit private corporate interests above public concerns. According to this logic, governments are left with 'no option' but to liberalise and deregulate if they are to maintain any control over the deployment of new technologies. Both US and British governments are currently engaged in a rethink of their cross-media ownership restrictions as a *necessary* response to the process of convergence, as if maintaining some sort of ceiling on market share would obviously lead to the collapse of the media industries.

It is this use of determinism, as a means of closing off alternatives, that Williams found particularly disingenuous in McLuhan's argument that 'the medium is the message'. Far from providing a radical vision of communications systems, the latter's fatalism played directly into the hands of the media establishment:

> It is hardly surprising that this conclusion has been welcomed by the 'media men' of the existing institutions. It gives the gloss of avantgarde theory to the crudest versions of their existing interests and practices, and assigns all their critics to pre-electronic irrelevance. Thus what began as pure formalism, and as speculation on human essence, ends as operative social theory and practice, in the heartland of the most dominative and aggressive communications institutions in the world.
>
> (Williams 1974: 128)

However, Williams was also contemptuous of those radical critics of new technology whose unthinking hostility towards communications innovations led them into a 'tacit alliance with the defenders of old privileged and paternalist institutions' (Williams 1985: 129). Williams may have been thinking of those in the British Labour Party who opposed the introduction of cable and satellite from a barely disguised position of anti-Americanism. The result of this was to adopt a conservative attitude towards new technology which coincided with an elitist defence of traditional broadcasting institutions, a position that blunted their justifiable criticisms of commercialisation and liberalisation. Williams attributed this 'cultural pessimism' to a deep-rooted 'minority culture' critique of 'mass communications' that he argued has been present in the early days of all new technologies, and a position that he himself had adopted in some of his early works. Now, however, 'as one after another of the stylish old institutions, which had supposed themselves permanently protected, is cut into by the imperatives of a harsher phase of the capitalist economy, it is no surprise that there is only a bewildered and outraged pessimism' (Williams 1985: 135). A socialist critique would have to avoid this kind of negative determinism and instead press for alternative structures for and uses of new technologies.

> 'Unforeseen uses and unforeseen effects' may qualify the 'original intention' of those developing the technology.
>
> (Williams 1974: 129)

While Williams emphasised the importance of 'purpose' and social intervention in the development of technologies, he rejected the idea that technologies would necessarily be used in the precise ways envisaged by the developers. Just as technological determinism was a misleading theory, so was the notion of a 'determined technology' (Williams 1974: 130). If technologies are social relationships, not static or predictable processes, they are therefore caught up in and shaped by social struggles. Technologies, in other words, have 'social complications' (Williams 1981: 230). For example, although political and religious authorities were keen for ordinary people to read the Bible in the nineteenth century for moral instruction, they 'overlooked the fact that there is no way of teaching a man to read the Bible which does not also enable him to read the radical press' (Williams 1981: 230). Private appropriation of the telephone and the photograph coexisted with their intended industrial uses in ways that led to 'wider and more varied personal and social contacts than had been possible within older and more settled communities' (1981: 233). Even in the case of television, a technology brought into being through corporate design, there are conscious attempts to transcend the limitations of 'mobile privatization'. Williams claims that there have been oppositional uses of television, for example the electronic town-meeting and the 'multi-screen play', where experimentation and participation are 'as much an effect as the more widely publicised and predicted passivity' (Williams 1974: 133).

This conception of the use of communicative activity to moderate the alienation and atomisation of industrial society flows from Williams' definition of a democratic communications system, originally sketched out in *Communications* (Williams 1967 [1962]). Here he counterposed what he sees as the essence of communicative activity, 'the sharing of human experience' (1967 [1962]: 33) to its actual uses in capitalist society, as a means of either money-making or propaganda. Democratic communications, on the other hand. depends on the 'right to transmit and the right to receive' (1967 [1962]: 128), independent of the market and the state. Calling for public ownership of all large-scale media systems, he proposed a public-service-oriented system that runs on the principle that 'the active contributors have control of their own means of expression' (1967 [1962]: 129). There is nothing intrinsic to the technologies that prevents these sorts of institutional forms from being realised apart from the present social organisation of society that debases communication and attempts to hide alternative forms of social structure. Twenty years after *Communications* (1962), Williams was still writing about the potential of new technologies to enhance civil society and deepen the connections between individuals and groups in opposition to corporations and the state.

Williams was by no means naive about the possibilities of challenging corporate, military and state control of communications systems simply through articulating alternative models of media. Although technologies are not preordained and immutable, neither are they 'undetermined'. Indeed, Williams argues that technologies are socially 'determined' in the sense that real determining factors – the distribution of power or of capital, social and physical inheritance, relations of scale and size between groups – set limits and exert pressures, but neither wholly control nor wholly predict the outcome of complex activity within or at these limits, and under or against these pressures.

(Williams 1974: 130)

Cable television is a perfect example of a technology that has been shaped by a range of conflicting forces: broadcasters, regulators, government, academics, engineers, corporate bosses, individual subscribers and community activists. Cable could have been introduced as a way of establishing a new and more direct relationship between broadcaster and viewer and to represent and involve minority groups independently of the definitions of advertisers and marketing experts. In reality, this approach has been marginal as compared to cable's capacity to offer extra streams of revenue to established media and telecommunications groups. Williams' argument is that it will be the best resourced groups, which under capitalism means corporations and the state, that 'determine' the most and least likely paths of development even if this 'determination' is up for constant challenge and rebuttal. As Williams puts it, 'whether the theory and practice can be changed will depend not on the fixed properties of the medium nor on the necessary character of its institutions, but on a continually renewable social action and struggle' (1974: 134). The future development of communication technologies is not preordained but subject to the outcome of wider battles over the shape and form of social life.

Williams' theory of determination as the 'setting of limits and the exertion of pressures' was designed to counter what he saw as the essentialism of the Marxist model of base and superstructure. This has its strengths but also its problems. Williams takes great care not to reduce complex technologies to the whims and desires of a few entrepreneurs or to the needs of abstract structures. He injects a sense of the importance of agency and intention into technological development but then assesses how these intentions are welcomed, modified or rejected in their eventual deployment by users. On the other hand, his notion of the *multiple* layers of determination and his emphasis on the continual interaction between different levels of social production and reproduction lead him to a certain elusiveness about which factors are determining and which are not. According to Terry Eagleton, Williams' concept of culture and communications as determining forces lacks the power of the Marxist formulation of base and superstructure where 'determinations are not symmetrical: that in the production of human society some activities are more fundamentally determining than others' (Eagleton 1989: 169). In his eagerness to demonstrate the materiality and productivity of what were often written off as being purely 'symbolic' processes, Williams tends to elide the significance of developments in the economic and the cultural spheres.

But is the Internet different?

William's insistence on recognising the social dimension of technology is particularly relevant today given the scale of the claims made about the Internet as a *transformative* medium. For example, how do we explain how the Internet was changed from a non-commercial instrument of mainly academic and military information exchange to a highly commercialised tool of mainly private and business transactions? For some, this is due simply to the power of the technology: 'The growth of the Net is not a fluke or a fad, but the consequence of unleashing the power of individual creativity. If it were an economy, it would be the triumph of the free market over central planning. In music, jazz over Bach. Democracy over dictatorship' (Anderson 1996: 97). This is a relatively typical view that the Internet, once 'liberated' from the restrictions of its public status, would inevitably thrive because of its innate tendencies towards competition, improvisation and decentralisation. However, this is not a disinterested position but a description of the Internet from a guide to digital technologies published by *The Economist*, a magazine dedicated to celebrating the dynamism of the free market. This picture of the Net was pushed by a number of theorists in the early 1990s, including Alvin Toffler, George Gilder and John Naisbitt, and taken up by many western politicians, most notably then house speaker Newt Gingrich and US vice-president Al Gore. The latter's speech at the 1994 International Telecommunications Union conference demonstrated a growing determination on the part of big business and its backers in the political field to co-opt the Net for its own purposes. Gore spoke in McLuhanesque terms of a Global Information Infrastructure (GII), promised that 'the distributed intelligence of the GII will spread participatory democracy' and predicted 'a new Athenian age of democracy forged in the fora the GII will create' (quoted in Leer 2000: 181–182).

This story of Internet technology as a natural ally of liberal democracy and the free market was used to justify the ensuing privatisation and commercialisation of cyberspace. It need not have followed this line of development, especially given its earlier public status. Other models or metaphors could have been adopted: an electronic public library, a public sphere independent of both state and market, a civic space leased to individuals and groups for public benefit and not private gain. Instead, the development of the Internet as a commercial space was the result of a decisive intervention by corporations and governments following neo-liberal ideas about the supposed benefits of consumerism and competition. This required the defeat of critics of the free market, a battle launched and inspired less by technological certainties than by a firmly held belief in the values of capitalism. George Gilder tackled his critics head on in 1995:

> Blinded by the robber-baron image assigned in U.S. history courses to the heroic builders of American capitalism, many critics see Bill Gates as a menacing monopolist. They mistake for greed the gargantuan tenacity of Microsoft as it struggles to assure the compatibility of its standard with tens of thousands of applications and peripherals over generations of dynamically changing technology . . . They see the Internet as another arena likely to be dominated by Microsoft and a few giant media companies, increasing the wealth of Wall Street at the expense of the stultified masses of consumers and opening an ever greater gap between the 'information rich' and the 'information poor'.
>
> (Gilder 1995)

While Gilder's critics have been proved right on all counts – witness the antitrust case against Microsoft, the growing concern about a 'digital divide' and the control over Internet traffic and content by a handful of corporations – the Net is still guided by free-market interests. This confirms Williams' argument that the shape that technologies assume owes a great deal more to the priorities of the most powerful interests in society than it does to any internal characteristics of the technology.

Technological determinism, however, remains a very powerful discourse in the attempt to construct a common-sense view of digital systems that they are innately competitive and democratising and thus unsuitable for public ownership or traditional forms of regulation. Let the market decide and watch consumers take control while bureaucracies and dictatorships crumble, goes the argument of an influential neo-liberal theorist like Francis Fukuyama, adviser to the US State Department.

> The newer information technologies are profoundly democratizing, because they don't reward economies of scale. They work best in decentralized, non-controlled societies. They're anti-authoritarian, because authoritarians control societies by their ability to control access to information. So if people can get information on their own

simply by dialing a computer, then we have ways of getting around hierarchies.

(Fukuyama 2000)

The consequence of this for Fukuyama is that the Internet must be adopted in a way that is favourable to the principles of liberal democracies and market economies. The problem is not only that Fukuyama hides the fact that there is a choice to be made about how to develop particular technologies but also that he appears to be wrong in his description of the Net's democratising tendencies. 'Far from hastening its own demise by allowing the Internet to penetrate its borders, an authoritarian state can actually utilize the Internet to its own benefit and increase its stability by engaging with the technology', argue the authors of a report for the Carnegie Endowment for International Peace (Kalathil and Boas 2001: 4). Furthermore, Fukuyama tends to exaggerate the decentralising tendencies of the Net. A recent major study of traffic on the web found that a tiny minority of highly commercial websites accounted for a significant amount of traffic: the top 0.1 per cent of all sites drew one-third of all user volume with the top 10 per cent of sites attracting 83 per cent of all 'hits'. The authors conclude that their evidence points to a 'signature of winner-takes-all markets' (Adamic and Huberman 2000). The sheer visibility of America Online (AOL) and Yahoo! (or of Wanadoo in France and T-Online in Germany) as gatekeepers to cyberspace points to a more complex account of new technologies than the deterministic one that Fukuyama and Gilder provide.

The point is that such determinism embodies a highly political account of the world and shrouds the real decisions about technological development in a veil of inevitability. A leading British exponent of 'new economy' determinism is Charles Leadbeater (2000) whose views on the weightless economy have been welcomed by British prime minister Tony Blair. Leadbeater contrasts the dynamism of contemporary innovation with what he sees as the more ponderous and accidental nature of innovation in the nineteenth century and urges us to embrace the economic value of knowledge today. 'In the knowledge and service economy products are weightless. They replicate like viruses at the speed of modern computers and communications systems' (Leadbeater 2000: 234). The theorising of the hegemony of the service economy and the reification of knowledge appears to be a disinterested, technologically informed practice but it has, of course, significant political connotations. For example, according to Anthony Giddens, the transformative power of digital technologies in a globalised world means that 'information and knowledge have now become media of production, displacing many kinds of manual work. Marx thought that the working class would bury capitalism but as it has turned out, capitalism has buried the working class' (quoted in Hutton and Giddens 2001: 22). For both writers, a belief in the productive and creative logic of new technologies underpins their defence of market relations in the information age and their belittling of generalised public ownership as appropriate only to a now disappearing industrial society.

The problem for such writers is that the performance of new technologies does not justify such an analysis. The production and distribution of tangible goods by groups of people who do not own or control the process stubbornly continues, despite the best efforts of neo-liberals to theorise these facts out of existence. Consider the downfall of the online grocer Webvan, described by the *Financial Times* as 'a symbol of the Internet's unlimited potential' (Edgecliffe-Johnson 2001). It collapsed, not because an e-grocer was a bad idea but because its business model – of building huge warehouses, eighteen times bigger than a normal supermarket, that would rely on machines to service customers' orders – required so much capital investment that it was simply not competitive with traditional retailers. Over $1 billion was wasted before investors realised that an evolutionary 'bricks and clicks' strategy was more relevant to the future of supermarket shopping. The idea that there is a technologically based 'new economy' unaffected by either the financial disciplines or even the reliance on labour of the 'old economy' (who, after all, was going to drive Webvan's delivery trucks?) appears to be a myth, painfully punctured in the recent slump in Internet-related enterprises.

Williams argued in *Towards 2000* that the practice of reading off social change from technological innovation 'is especially misleading in descriptions and predictions of a "post-industrial" society. For in the end it is impossible to understand the industrial revolution in any of its phases, including the most recent and most imminent, by reference to the changes in the forces of production alone' (Williams 1985: 84). Only by locating technologies inside existing social relations, thereby appreciating some of the conflicts and contradictions in technological development, can we start to grasp the possibilities and the limitations of particular innovations. Following Williams, we can argue that the Internet is neither the empowering, decentralised technology that the Negropontes and Gilders would have us believe, nor the instrument of isolation and atomisation that some critics allege it to be. Williams' theory of 'mobile privatization' helps us to further situate the Internet today as a technology that connects the ever-increasing flows of social mobility (of migration, commuting, capital flows and tourism) to the privatised enclave of the family home and the screen of the individual user. The Internet is neither a determined nor a determining technology and its future depends on the result of the struggles that take place over both immediate questions – such as copyright and privacy in cyberspace – and more profound ones concerning the growing market orientation and corporate control of contemporary social life.

Conclusion

Williams' work is absolutely essential for anyone seeking to grasp the dynamics of technological innovation in a society in which technology is increasingly both deified *and* reified. Williams helps to remind us that technological development is neither a magical solution to declining productivity and growing inequality nor an autonomous process over which humans have little or no control. His books and articles provide a refreshing account of the interests behind technological development in contrast with the neo-liberal efforts to 'market' new technologies by presenting them as both desirable and inevitable. By stressing the fundamentally social nature of technologies, he illuminates the social and economic contexts in which innovation takes place and assesses the impact of technologies on the societies into which they are introduced in a dialectical, not mechanical, fashion. He also challenges the negative determinism of those critics who *dismiss* new technologies simply on the basis that they necessarily reflect the interests of the most powerful in society. Instead, Williams points out the contingent nature of technological development. He paints a complex picture of innovation as a process marked by the priorities of dominant groups that limits the full range of technological possibilities but also as a process swayed by the social struggles that envelop all societies.

Williams' commitment to democratic communications and his recognition of the possibilities of new technologies under a different social system offer a vital challenge to the free-market consensus about the new media today. His critique of technological determinism and his emphasis on the sociality of technologies is a timely counterbalance to the voices (perhaps less shrill than they used to be) that profess digital technologies to be the embodiment of competition and liberal democracy. Finally, Williams was an intellectual and an activist who sought to ground his understanding of socialist theory in political practice, not on an international lecture circuit that is currently littered with academics selling themselves and the information revolution. While the London School of Economics' Ian Angell, described as 'Europe's leading IT guru and visionary', is yours for between $30,000 and $50,000 (Leading Authorities

2001), a wiser investment would be a paperback copy of *Television: Technology and Cultural Form*.

References

Adamic, L. and Huberman, B. (2000) 'The nature of markets on the World Wide Web', *Quarterly Journal of Electronic Commerce* 1: 5–12. Available: ftp://ftp.parc.xerox.com/pub/dynamics/webmarkets.pdf (25 July 2001).

Anderson, C. (1996) 'The Internet' in The Economist, *Going Digital: How New Technology is Changing our Lives*, London: Economist Books.

Eagleton, T. (ed.) (1989) *Raymond Williams: Critical Perspectives*, Cambridge: Polity Press.

Edgecliffe-Johnson, A. (2001) 'Webvan joins list of high-profile net flops', *Financial Times* (US edition), 10 July: 1.

Fukuyama, F. (2000) Transcript of a programme on 'Will the Internet Change Humanity?', Closer to Truth, Show 102. Available: http://www.closertotruth.com/topics/technology/society/102/102transcript.html (13 August 2001).

Giddens, A. (1999) *Runaway World*, London: Profile Books.

Gilder, G. (1995) 'Angst and awe on the Internet', *Forbes ASAP* 4 December. Available: http://www.seas.upenn.edu:8080/~gaj1/angstgg.html (13 August 2001).

Hutton, W. and Giddens, A. (eds) (2001) *On the Edge: Living with Global Capitalism*, London: Vintage.

Kalathil, S. and Boas, T. (2001) *The Internet and State Control in Authoritarian Regimes: China, Cuba, and the Counterrevolution*, Washington, DC: Carnegie Endowment for International Peace. Available: http://www.ceip.org/files/pdf/21KalathilBoas.pdf (15 August 2001).

Leadbeater, C. (2000) *Living on Thin Air: The New Economy*, revised edition, Harmondsworth: Penguin.

Leading Authorities, Inc. (2001) Biography of Ian Angell. Available: http://www.leadingauthorities.com/search/biography.htm/s/10443.htm (18 August 2001).

Leer, A. (2000) *Welcome to the Wired World*, an ft.com book, Harlow: Pearson Education.

McLuhan, M. (1998) [1964] *Understanding Media: The Extensions of Man*, London: MIT Press.

Negroponte, N. (1998) 'Beyond digital', *Wired*, 6 December. Available: http://www.media.mit.edu/~nicholas/Wired/WIRED6-16.html (13 August 2001).

Sawhney, H. (1996) 'Information superhighway: metaphors as midwives', *Media, Culture and Society* 18(2): 291–314.

Williams, R. (1958) *Culture and Society, 1780–1950*, London: Chatto & Windus.

Williams, R. (1961) *The Long Revolution*, London: Chatto & Windus.

Williams, R. (1967) [1962] *Communications*, revised edition, New York: Barnes & Noble.

Williams, R. (1974) *Television: Technology and Cultural Form*, London: Fontana.
Williams, R. (1981) 'Communication technologies and social institutions' in R. Williams (ed.) *Contact: Human Communication and its History*, London: Thames & Hudson.
Williams, R. (1985) *Towards 2000*, Harmondsworth: Penguin.

Index

*The subjects of specific chapters only appear in this index when mentioned in other chapters.